Responding to the Livestock Revolution

The role of globilisation and implications for povery alleviation

Responding to the Livestock Revolution

Edited by

E Owen, T Smith, MA Steele, S Anderson, AJ Duncan, M Herrero, JD Leaver, CK Reynolds, JI Richards, JC Ku-Vera

First published by Nottingham University Press

This reissued original edition published 2023 by
5m Books Ltd www.5mbooks.com

British Library Cataloguing in Publication Data

Responding to the Livestock Revolution: the role of globilisation and implications for povery alleviation
i Owen, E., ii Smith, T., iii Steele, M.A., iv Anderson, S., v Duncan, A.J., vi Herrero, M., vii Lever, J.D., viii Reynolds, C.K., ix Richards, J.I., x Ku-Vera, J.C.

ISBN 9781789183047

Disclaimer

Every reasonable effort has been made to ensure that the material in this book is true, correct, complete and appropriate at the time of writing. Nevertheless, the publishers and authors do not accept responsibility for any omission or error, or for any injury, damage, loss or financial consequences arising from the use of the book.

Typeset by Nottingham University Press, Nottingham
Printed and bound by PODWW, UK

CONTENTS

areas of the different countries (Devendra and Thomas, 2002). In India, meat production has increased considerably during the last few years and is projected to continue to grow. Small ruminants and chickens are the main source of meat in India.

Relatively more attention will continue to be given to those mixed crop-animal systems that involve annual crops, because of their importance as food staples. However, the decreased availability of arable land in many areas, as human population densities increase, could encourage further integration of ruminants with tree crops. In regions such as Southeast Asia and the more humid areas of South Asia, millions of hectares of land under tree crops (coconut, oil palm, fruits and rubber) are potentially available for this purpose (Devendra, 2002). Already, associations of tree crops and animals are established practice in many developing countries. The expansion and intensification of these systems is a realistic objective given the extent of farmer experience, the periodic collapse of world prices for plantation commodities and the projected demands for livestock products. Contrary to popular belief, a significant area of tree crop plantations is in the hands of small-scale farmers as a consequence of land reform in developing countries. In irrigated areas of western India excellent integration of sugarcane and milk production has emerged in recent decades (Rangnekar, 1988).

Constraints to animal production

Most of the production gains in ruminants in developing countries have come from an expansion of the livestock populations, rather than through significant increases in gain/animal. However, as systems intensify and the area of grazing land remains fixed, or most likely reduced by crop encroachment and degradation, increasing animal numbers is not a sustainable long-term option in many countries. Productivity improvements in milk and meat are projected (FAO, 2002), which will allow increased production with lower growth in animal numbers in some countries.

If mixed farming systems are to contribute significantly to the projected demands for animal products and improve livelihoods, then productivity/animal will need to continue to increase. This will require the use of appropriate technologies, along with an expanded use of inputs, supported by suitable policies and better infrastructure. Technical, institutional, social and economic constraints to increased animal production will need to be identified and addressed. These constraints in developing countries are well known and have been discussed at depth in Gardiner and Devendra (1995).

Animal genetic resources

Genotypes are both a constraint and a benefit to improved productivity, particularly in the rural-based small-scale production systems (Vercoe *et al.*, 1997). On the one hand, a poor genotype imposes limits on the productivity that is being achieved and on the potential for increases from new technologies. On the other hand, developing countries are rich in domestic livestock biodiversity, and superior genotypes with resistance to environmental stresses exist within the indigenous breeds of various species. Additionally, exotic breeds are available for rapidly increasing production through crossbreeding and the exploitation of hybrid vigour. However, animal breeding programmes in most developing countries have tended to neglect the characterisation and improvement of valuable indigenous breeds that are generally most associated with small-scale farming systems. Intensification is leading to a reliance on a narrow range of improved, high-yielding breeds in many species.

Dairy production, for example, has achieved major attention in developing countries, and enormous resources have been invested in crossbreeding of indigenous stock with exotic animals. However, dairy development has had varying degrees of success. Spectacular increases in milk yields have been registered in, for example, parts of India. Nevertheless, there have been major problems in the practical implementation of these programmes (Thomas *et al.*, 2002), resulting in limited impact at farm level. Through studies on the outcome of livestock development programmes in India, it has become increasingly clear that improved animals with high production potential are only beneficial in areas where farmers have adequate resources, and are not a panacea across all production systems. In Zimbabwe, Smith *et al.* (2002) concluded that, although on-station research shows advantages in the use of cross-bred dairy cattle, for resource-poor farmers crossbreeding of indigenous animals with exotics to increase production is not a recommended option. The authors further recommend that indigenous breeds should be subjected to selection for specific production traits.

The concentration of massive resources in this way has often diverted attention from the wider use of other species that contribute to production. These include the buffalo, small ruminants and ducks. In India and Pakistan, buffalo contribute more to national milk production than cattle (Devendra *et al.*, 2000), whilst small ruminants are the main source of meat for the poor in Asia, sub-Saharan Africa and in the West Asia-North Africa region.

Feed resources

It is generally acknowledged that poor nutrition at farm level is one of the greatest technical constraints to increased animal productivity. A wide range of feed resources is available in theory and include native grasses, weeds, crop residues, agro-industrial by-products, cultivated forages and tree foliage (Renard, 1997). However, in the small-scale mixed farming systems, the proportion contributed to diets by cultivated forages is relatively small and still remains to be exploited. Even in the irrigated systems, in countries such as India and Pakistan, only 3-15% of the total arable area is used for cultivated forages (Devendra *et al.*, 2000). Millions of hectares of improved grasses have been sown in grassland-based systems in South America, but many of these pastures have not been sustainable because of fertility and pest problems. Additionally, forage legumes have not been sown widely, despite the availability of species adapted to the acid, infertile soils of the region. Riveros *et al.* (1993), Carter (1995) and Thomas and Sumberg (1995) have all discussed the reasons for poor adoption of improved pasture species. In those countries where areas of native grasslands are in decline, because of the encroachment of arable farming, crop residues (principally cereals and pulses) are becoming increasingly important for ruminants. Of equal importance are the by-products of oilseed processing, grain milling and other agro-industrial activities. However, these agro-industrial by-products tend to become available in the peri-urban/urban centres, not in rural areas, and are more associated with specialised non-ruminant systems. In India, agro-industrial by-products and non-conventional feed resources contribute <10% to the ruminant feed supply (Devendra *et al.*, 2000).

Ironically, in the humid tropics, ruminant feed resources are often in excess of ruminant requirements and remain under-utilised. Devendra (1995) has calculated that, in Malaysia, feed availability from a range of feeds is about four times greater than requirements. Similarly, in the Philippines, available feeds can support six times the existing ruminant populations. In these situations increased utilisation of available feeds is restricted currently by inadequate animal numbers, a lack of intensification of the production systems and poor technology delivery and use at farm level.

In the drier tropics animal feed deficits are common. For example, in India, Singh *et al.* (1997) indicated a substantial gap between supply and demand of animal feed of about 31%. Most critical was the deficit of digestible crude protein, which amounted to 58%, showing a severe imbalance in the ratio of poor-quality forages to good feeds. The challenge is to increase the availability of feeds, both in terms of quantity and quality (particularly protein content). According to Vercoe *et al.* (1997) there are two components to this; the first is to improve the seasonal and annual supply of nutrients, and the second is to reduce the competition for available feeds. Even in the dry tropics, feeds can be poorly utilised, wasted or not used at all. The second component involves a reduction in the numbers of animals that are being kept on maintenance or near maintenance rations, so that the remainder can be fed at higher levels.

The projected expansion of ruminant populations (FAO, 1993) will result in marked increases in the amounts of energy and protein required from forages and other feed resources. Additionally, a rise in the numbers of non-ruminants, and intensification away from scavenger systems, will increase the demands for cereals and oilseed proteins. Some three-fifths of world consumption of coarse grains are used by livestock (FAO, 2002). Consumption has been rising fast and, in future, might exceed that for rice or wheat, in line with growth of the livestock sub-sector. Developing countries will account for a rising share of world production, from <50% at present to just under three-fifths by 2030 (FAO, 2002). Average demand for roots, tubers and plantains is also projected to rise in developing countries, with sweet potato and the Irish potato becoming important as livestock feed (FAO, 2002).

Estimates for ruminants and equines in sub-Saharan Africa (Winrock, 1992) indicate that annual forage consumption will more than double by 2025, and that the needed output of livestock products will not be achieved from current forage production. The situation for protein is even more acute. The rangelands of the arid zone are near maximum production, and opportunities for increased output there are minimal. In the tropical highlands, there are good opportunities for farmers to raise production by increased use of new technologies and inputs. The higher rainfall areas of the semi-arid zone and the sub-humid zone can also provide opportunities to produce the additional feed required for expanded ruminant and equine populations. For non-ruminants, feed grain requirements are expected to increase 10-fold by 2025 and oilseed requirements by even more. According to Winrock (1992), increases in grain and root crop production in sub-Saharan Africa of 4%/year will make the goal of meeting animal feed requirements for increased production achievable.

The relationship between feed resources and livestock production has always been a subject of debate in India since a sharp rise in the production of milk, meat and eggs has been achieved, along with an increase in livestock populations, even though feed deficits have been reported over many decades. A recent report on feed/forage resources in India indicates that the situation is not that serious, but there is a need for the development of proper methodologies to provide realistic estimates of these resources (Ramchandra *et al.*, 2001).

Animal health and diseases

Diseases are a major constraint to livestock production in developing countries. Animal health issues are barriers to trade in livestock and their products, whilst specific diseases decrease production and increase morbidity and mortality at farm level. A wide range of infectious and other diseases caused by bacteria, parasites and viruses is prevalent in the Third World, and the major ones are often common to all countries in a particular region (Devendra *et al.*, 1997; Devendra *et al.*, 2000). Additionally, there are reproductive, metabolic and mineral disorders in ruminants, and toxicosis in animals raised in peri-urban and urban areas as a result of environmental pollution and feed adulteration (Thomas, 1999).

Mortality, morbidity and economic losses can be high. For example, in Bangladesh, foot and mouth disease outbreaks occur at least twice annually. In every outbreak, around 50% of ruminants are affected, and the annual economic loss is estimated at US $125 million (Devendra *et al.*, 2000). Anthrax, black quarter and haemorrhagic septicaemia outbreaks occur sporadically, resulting in up to 40% mortality in cattle and an annual loss estimated at US $2.3 million. In India, the direct loss to foot and mouth disease is estimated at more than US $430 million annually. Additionally, animal products from such regions are not accepted by other foot and mouth-free countries, reducing further the export potential of the sub-sector. In Pakistan, the overall annual losses due to diseases are estimated at US $200 million.

Often, disease diagnosis and monitoring systems are weak in developing countries, as is the understanding of the epidemiology of the important diseases. Thus, there is relatively little information on the incidence of diseases, their geographical/seasonal distribution or dynamics. In India, for example, there is no organised system for disease reporting (Devendra *et al.*, 2000). The existing system is passive and does not report the confirmation of a disease. A lack of reliable data is interpreted as the absence of the disease. An All-India project has now been launched by the Indian Council for Agricultural Research (ICAR) to monitor and survey animal diseases, and has already made an important contribution to the development of simulated animal disease-forecasting models.

Vaccines are produced to some degree in many countries, but they are often inadequate to meet national needs and there are problems of quality control. Biotechnology has yet to play an important role in vaccine development in the Third World. Veterinary delivery systems are also inadequate in many of the countries, and farmers have little confidence in the state-run services. Access to these services is very difficult in remote rural areas, and often farmers have to resort to help from traditional healers and the use of ethno-veterinary medicine to treat animals. Other reasons for the renewed interest in ethno-veterinary medicine are the costs of treatment, the problem of residues in products and the development of resistance in disease organisms, particularly as a result of the indiscriminate use of antibiotics (McCorkle *et al.*, 1999). Whilst claims of success are often anecdotal, there is a large body of literature indicating a need for re-examination of the role of ethno-veterinary medicine in smallholder production systems. Conroy and Thakur (2002), in India, provide an example of the use of *Mucuna pruriens* as an anthelminthic for goats. The idea for its use came from members of a local caste specialising in buffalo keeping.

Other factors

By comparison to the major crops, the attention given to the livestock sub-sector is minimal, and in no way reflects the contribution that animals make to the agricultural economy. National policies are often designed to promote crop production, and a wide variety of inappropriate policies have seriously hampered rather than stimulated the development of animal industries. Modernising the livestock sub-sector and providing the flexibility for it to respond adequately to an increased global demand for animal products requires a change of attitude and appropriate development policies. This should be reflected in appropriate allocations of funding and improved institutional facilities. The worst thing that could be done is for agencies to cease the public investments that facilitate the development of sustainable, small-scale systems of market-orientated livestock production.

The price farmers receive for products can be a disincentive to adopting new technologies or using agricultural inputs. Credit for the purchase of livestock, particularly large ruminants, is often unavailable outside of externally-funded development projects or difficult to access due to bureaucracy. In some communities, poor farmers and women have to begin with the purchase of small stock in order to generate finance for acquiring larger animals. In Central America, it is often pigs, and in sub-Saharan Africa and Asia, small ruminants. Even for small stock credit is often necessary, and people are forced to borrow money at high interest rates from unscrupulous traditional money lenders. The example set by credit institutions like the Grameen Bank in Bangladesh needs to be followed. This famous institution has provided about 50% of its loans and approximately 25% of its loan fund to the landless and smallholders, principally women, for livestock-related activities (Devendra *et al.*, 2000). Dairy cattle, beef fattening and goat and poultry rearing are included for extending credit facilities. In India, the formation of a large number of self-help groups in rural areas,

To compete effectively with industrial systems, small-scale livestock producers need improved feed and feeding technologies. Advantage needs to be taken, in the case of ruminants, of their capacity to utilise feed resources inedible to humans. Enhancing the supply of nutrients to livestock can be achieved by improving the availability of local foodstuffs (new forages, agricultural by-products, and non-conventional feed resources), improving the efficiency of utilisation of feeds (rumen manipulation, supplementation, and treatment/processing) and increasing feed imports. Feed resources for livestock rearing are often wasted, under-utilised or not used at all, even in the most marginal environments. Furthermore, *ex post* economic assessment of planted forages in West Africa by Elbasha *et al.* (1999) has shown that the lag period associated with the diffusion of this technology is considerable-at least 15 years-and may be longer than anticipated.

A great deal is known about animal diseases and their treatment. Furthermore, biotechnology is providing new opportunities for disease control. As mentioned previously, a major problem in developing countries is the delivery of animal health services which is primarily a state activity. The supply and quality of these services, outside of co-operatives and externally-funded projects, has been poor. To increase the availability of, and access to, animal health services, there should be an enhanced role for the private sector (de Haan, 1995). However, the privatisation of livestock services has to be undertaken with caution, since there is a danger that resource-poor livestock owners may be left out of the process. A gradual change in the system would be the most appropriate approach. Alternative ways for improving the accessibility to livestock services are being tried in many developing countries; the development of teams of para-veterinarians by non-government organisations and co-operatives in Bangladesh, India, Kenya and Nepal, and the decentralisation of services in Indonesia are examples. Access to livestock services for dairy animals in India has improved considerably through the network of dairy co-operatives. The state should focus on areas such as sanitary and border controls, quarantine operations and extension, whilst the private sector should take over clinical treatment of animals and non-compulsory vaccinations. New vaccine technologies based on robust, one-shot and oral applications are a high priority (Vercoe *et al.*, 1997).

There is a large variation between developing countries in the quantity and quality of the infrastructure that supports livestock research (Devendra *et al.*, 1997; Devendra *et al.*, 2000). In general, there is a substantial need to improve extension services to small farmers involved in animal production. Livestock extension has been marginalised by major interest groups (crop-based extension and animal health services), and by a lack of a clear understanding of the needs of livestock farmers (Matthewman and Morton, 1997). Crop production needs and animal health problems are more easily addressed than livestock production needs. Improvements need to focus on the nature of integrated farming systems, dissemination pathways for information and the training needs of producers, if the best use is to be made of the information provided. Advances in information technology are being channelled through the National Technology Project in a few states in India, to cater for the needs of the farming community for information. The Project takes an integrated farming systems approach, and makes use of Geographical Information Systems for providing the desired knowledge to farmers through booths set up in rural areas. The National Dairy Development Board is developing a National Information Network for co-operatives to cater for their information needs.

Responding to the challenge

A survey of the conventional literature indicated a paucity of quantitative information on the impacts of new technologies (and their sustainability) on the livelihoods of resource-poor livestock farmers in developing countries. Nevertheless, the following are a few examples of how small-scale farmers have increased the productivity of their animals and improved their livelihoods through the uptake of new knowledge and the use of inputs.

The three-strata forage system in Indonesia

In many countries, multi-purpose trees and shrubs have been introduced into small-scale agricultural systems to increase the availability and nutritive value of the feed supply. On Bali, in Indonesia, the Three-Strata Forage system (TSFS) has been developed. Forages (grasses and herbaceous legumes), shrubs and trees are grown on an area of 0.25 ha, and integrated with crops such as maize, soyabeans and cassava. The effects of TSFS on cattle productivity and livelihoods on small mixed farms compared, over nine years, with those not utilising the new technology are shown in Table 1.2.

Forage yields were increased by 91% and the feed contained 13% more crude protein due to the legumes. The increased forage production allowed higher stocking rates to be maintained and resulted in liveweight gains in Bali steers equivalent to 375 kg/ha/year. On smallholdings not practising the TSFS, animal gains were only 122 kg/ha/year. Additionally, there was almost a five-fold increase in firewood production. Soil erosion was reduced by 57%, and soil organic matter increased by 11%. As a result, farm incomes (returns above variable costs) increased by 30%, and income per family by 40%.

Table 1.2. Productivity over nine years on small-scale mixed farms with the introduction of the Three-Strata Forage System.

Attributes	NTSFS*	TSFS	% Difference
Forage yield (kg DM/year)	573	1096	91
Cattle liveweight gain in three years (kg)	166	186	11
Carrying capacity (kg liveweight/year)	200	300	50
Time to reach market weight (375kg) (years)	4.3	3.8	-12
Farm income (Rp/year)	427,800	554,200	30
Farmer income (**Rp/family/year)	1,763,000	2,464,400	40

Source: Modified from Paris (2002).
*NTSFS is non-Three-Strata Forage System; TSFS is Three-Strata Forage System.
**Rp=Rupiah (Indonesian currency).

Improved forages in Costa Rica

The Centro Internacional de Agricultura Tropical (CIAT), in association with the Tropileche Consortium, has been involved with the intensification of production on small, dual-purpose dairy farms in Central America (CIAT-MAG, unpublished data). New forage-based technologies have been introduced for pasture improvement on hillsides. The impact of improved forages on animal productivity and incomes, in the dry tropics of the Central Pacific region of Costa Rica, is illustrated in Table 1.3.

Table 1.3. Productivity on a small-scale mixed farm following the introduction of improved forages in Costa Rica.

Parameter	Before 1996	Now 2000	Change (%)
Improved forages (ha)	4	10	150
Milking cows (number)	12	16	33
Milk yield (kg/cow/day)	3.3	5.1	55
Income (US$/day)	12	25	108
Area under forest (ha)	4	7	75

Source: CIAT/MAG unpublished data.

More milk was produced on a smaller area and farm incomes were doubled. As a result, land was released to serve as areas for timber production and the protection of water sources.

Small animal species in Bolivia

Paterson and Rojas (2002) have highlighted the importance of small animal species (chickens, ducks, pigs, hair sheep, Guinea pigs) to the livelihoods of small-scale farmers in the forest margins of the tropical Bolivian provinces of Sara and Ichilo, some 100 km north-west of the city of Santa Cruz. Their studies have shown that the implementation of simple, cheap, available recommendations for vaccination and parasite control, coupled with the provision of rustic, locally-made shelters, reduced losses of young animals and increased the productivity and profitability of the holdings.

The increases in productivity with improved management over traditional management were 97.7% for ducks, 62.9% for sheep, 29.5% for chickens, 16.0% for pigs, and 4.4% for Guinea pigs. Mortality rates under improved management ranged from 11.0% to 28.3%, compared to a range of 10.5% to 54.1% under traditional management. The interventions resulted in increased annual incomes of about US$213 and US$207 for lowland and highland families, respectively.

Smallholder dairy production in Kenya

Dairy production and marketing dominates the livestock contribution to household economies in rural areas with medium/high agricultural potential in Kenya (Muriuki *et al.*, 2002). Smallholder crop-dairy systems in these areas support up to 75% of the rural population of Kenya and a large proportion of the ruminant livestock. Milk is a very important source of income and food for an estimated 625,000 small-farmer households, and for those involved in marketing of the product; in total, some 25% of all households. About 4-6 litres/household of milk are sold daily, and two litres kept for home consumption.

Surveys conducted by Muriuki *et al.* (2002) have indicated that the adoption of dairy production and new technology has played a key role in the intensification of small-scale farming systems. Intensification has led to an increase in productivity in rural households that are often faced with sub-division of their farms, due to land inheritance systems, and poverty through the lack of non-agricultural job opportunities. Also, the introduction and marketing of improved forages has provided a new source of income. For example, in Kiambu District, the value of Napier grass (*Pennisetum purpureum*) and maize forage traded in 1996 was equivalent to US$5 million. Additionally, the value of manure may equal up to 30% of the value of milk produced. This benefit is captured on farm through the increased value of crop production, following the application of manure to the crops. Small-scale milk marketing hires more labour than non-dairy households, and generates 2-3 times the minimum daily wage.

Integration of dairy goats and forages on sloping land in the Philippines

Sloping areas in the Philippines have suffered from severe environmental degradation. On Mindanao, sloping agricultural land technology (SALT) has been developed to minimise soil erosion, restore soil fertility, sustain food production and generate regular and adequate incomes for the people residing in these areas. Amongst the SALT variants (Laquihon *et al.*, 1997), SALT-2 integrated crops and livestock. Half of each farm was planted to a forage garden and a livestock barn, and the other 50% was planted to mainly perennial field crops and contour hedgerows. Dairy goats were introduced onto the farms, 6-8 months after planting the forages, at a stocking rate of one buck and 12 does/0.5 ha. The goats were fed a concentrate mixture, and the forages consisted of a number of multi-purpose trees. SALT-2 produced 4,600 litres of goat milk annually for sale and home consumption.

Table 1.4. Cost and returns analysis (US$) of SALT-2 farms in the Philippines.

Items	*1991*	*1992*	*1993*
Total cash income (Cropping; sale of kids; sale of meat and milk)	3505.94	3635.37	3360.13
Total non-cash income (Replacement of stock; added value to stock; manure)	1416.80	1516.80	1556.80
Total gross returns	5022.74	5152.17	4862.93
Total costs	4156.96	3212.32	3607.87
Net profit (Total gross returns-Total costs)	865.78	1939.85	1255.06
Net cash returns (Total cash income-Total cash costs)	1103.30	2253.72	1713.11
Returns on investment	20.8%	60.39%	34.78%
Benefit-cost ratio	1.20	1.60	1.35

Source: Modified from Paris (2002).

Table 1.4 shows the economic benefits of SALT-2. The mean annual income for the three years was US$1,690 per 0.5 ha, with average returns on investment of 38.7%. The mean annual income was 14 times higher than that generally in the Philippine uplands. Various SALT farms occupy 5,000 hectares in the country. The technology has become the basis for government extension efforts to develop some 24 million hectares of sloping lands. The technology has also been extended to India, Laos and Sri Lanka.

Dairy production in India

Since 1971, several special programmes have been established to improve livestock production and the livelihoods of resource-poor farming families. One of the most famous, "Operation Flood," launched by the National Dairy Development Board, has provided much-needed financial, technical and organisational support for dairy production through a network of co-operatives. The positive impacts of "Operation Flood" have been reported by Shukla *et al.* (1995), Sampark (1997) and Shukla and Brahmankar (1999). Families from all livelihood groups involved in improved dairy production have benefited, and markedly higher incomes have been realised compared to those families not producing milk. More importantly, the proportion of dairy income to family income is highest for the landless and decreases with size of holding.

In the state of Uttar Pradesh, improved dairy production has generated more than 200 hours of employment per animal, helped to bring families above the poverty line, and contributed 53% to family income (Singh, 1995). In Kerala state, the incomes of families below the poverty line have increased by 49%, and improved dairy production has generated considerable employment (Krishnan, 1997). In Haryana state, improved dairy production resulted in 37 additional man-equivalent days of work and increased incomes substantially. Dairy production contributed about 33% to total family income (Singh *et al.*, 1998).

In Haryana state, Mathur and Singh (2002) have reported the impacts, after 20 years, of integrated dairy and crop production technologies on smallholder production. The original aim was to increase income levels and employment in 40 villages. The technology package included genetic improvement of dairy cattle and buffalo, improved nutrition, including the introduction of forage crops, and disease control. Crop technologies included new cultivars and the uses of inputs such as fertiliser and pesticides. Comparing the project area with the non-project area, the new technologies had a positive impact on milk production. Milk yields in crossbreds, local cows and buffalo were increased by 17%, 34% and 39%, respectively. Yields of rice and wheat were increased by 37% and 8%, respectively. Labour employment increased in all households. The authors concluded that the adoption of new technologies had a positive and significant effect on farm incomes.

Specialisation and involution

Steinfeld *et al.* (1997) have indicated two possible scenarios for the future of crop-animal systems in developing countries. One is where market forces and technological requirements force the systems to grow in unit size and to specialise, presenting fewer opportunities for on-farm crop and animal integration and the creation of environmental problems. The second possibility, as a result of continuing human population pressures in agro-ecological zones such as the tropical highlands, leads to a decrease in farm size to the point where the system collapses (involution). Large ruminants can no longer be maintained on the farm, the nutrient and farm power balance runs into a widening deficit and natural resources degrade. With the disappearance of the resource-enhancing role of animals, the environmental balance can be disrupted. Where involution might take place, this will need to be fought through accelerated technology uptake.

Industrialisation of production

There is concern that the main beneficiaries of the "Livestock Revolution" will be the large-scale urban producers. Currently, industrial systems produce 40% of the pork and 80% of the poultry meat (FAO, 2002). Intensive non-ruminant production in urban areas is expected to grow because of the inability of expanded red meat production to completely meet demands. Most of the small-scale pig and poultry production takes place in scavenger systems or in peri-urban units in developing countries. Under this system of extensive production, growth is slow and productivity low. There is potential for expanding indigenous non-ruminant production by smallholders, if local feed production can be increased. However, there is already a trend away from these systems towards industrial systems, particularly for pork, in Southeast and East Asia. This current trend towards industrial production to supply supermarkets could pose a threat to the rural poor who depend on livestock, and they will lose the one growing market that they presently supply. Policy measures that will help the poor enter and stay in an expanded market are needed. These measures include low-cost credit, technical support and better access to markets through improved infrastructure and institutions.

Additionally, a rise in the production of non-ruminant products may increase the price of cereals for the poor.

Public health concerns

Animals have long been recognised as agents of human diseases. Growing concentrations of animals and people in the major cities of developing countries can increase the incidence of zoonotic diseases, which can threaten livestock producers and consumers and reduce export potential. Some 75% of the most important zoonoses are associated with domestic animals (Thomas, 1999). These are particularly relevant to the industrial systems, but also present a threat to large concentrations of animals in the urban milk colonies of South Asia. Here, animals are often owned by groups of small producers. The control of zoonoses depends on the type under consideration, and may involve mass immunisation, slaughter of infected animals, environmental disinfection and quarantine. Supportive measures to control the spread of zoonoses through fluid media such as milk include treatments such as pasteurisation. The development of disease-free zones has also facilitated the export of meat.

In many developing countries, animals are subjected to rigorous controls to protect consumers and livestock producers from diseases. However, often, the large public-sector bodies created for this purpose have suffered fiscal crises that have limited the effectiveness of the services. On the other hand, countries such as India have made significant advances in the area of animal health (Devendra *et al.*, 2000).

The "Livestock Revolution" also raises other major public health concerns related to the transfer of pathogens and the build-up of contaminants such as carcinogenic mycotoxins (Coker, 1999) and pesticides (Thomas, 1999) in animal products. As the consumption of livestock products increases in developing countries, food safety risks become more prevalent. Furthermore, food contamination can affect the export potential for animal products and feeds from developing to developed countries. According to Nair (1998), in India, the bulk of pesticide residues (>70%) enter the human body through dairy products. Urban livestock populations are also exposed to heavy metal pollutants, notably lead, that can cause toxicosis (Thomas, 1999). This heavy metal has no known biological function, but every organ and system in the body can be affected. Therefore, the potential for accumulation in animal products is high. Again, industrial systems use antibiotics on a large scale, which can lead to the development of resistant bacteria including those causing human diseases.

In sub-Saharan Africa, agricultural productivity is threatened by the high AIDS mortality of the rural labour force. African agriculture is highly labour intensive, and employs two-thirds of the labour force (World Bank, 2000). The bulk of the human population of the nine countries worst hit by AIDS mortality depends on subsistence agriculture for survival (Hunter *et al.*, 1993). Additionally, the economies of most of these countries depend on agricultural exports to provide capital for development. At the end of 1998, an estimated 21.5 million men and women and one million children were living with HIV in sub-Saharan Africa (White and Robinson, 2000). The epidemic was originally understood principally to affect urban areas. However, HIV/AIDS prevalence continues to increase in the rural zones. Rural HIV seroprevalence rates range from zero to as high as 50% in some areas. If current infection rates were only 10%, the adult labour force would be reduced by 25% by 2010, and the average household would lose one or two adults (Norse and Stover, 1990, cited by Hunter *et al.*, 1993). In a survey in two agriculturally-important districts of Uganda, Hunter *et al.* (1993) found that of the 62% of the deaths attributed to AIDS and another long unidentified illness, some 74% were in the productive age group of 15 to 55 year-olds. Although AIDS mortality in these districts contributed, along with other pre-existing conditions, to a decline in agricultural productivity and household incomes, there appeared to be no direct effect on the decline in livestock production. Nevertheless, loss of labour and income would be expected to affect livestock as well as crop production. Furthermore, any increase in AIDS mortality does reinforce the vulnerabilities amongst farmers that are created by other factors (Hunter *et al.*, 1993). Wider coverage of HIV/AIDS and rural livelihoods in sub-Saharan Africa, and the relevance of the disease in relation to current poverty-alleviation initiatives is given by White and Robinson (2000).

Macroeconomic, international trade and sectoral policies

In developing countries, demand for animal products will grow faster than production (FAO, 2002). For meat products demand will rise steeply, whilst for milk and dairy products the rise will be less steep, but still considerable. Thus, a growing deficit will need to be satisfied by imports, which might suppress the development of the local production systems. There is particular concern over the EU and the USA, where domestic policy regarding subsidised production of livestock products and livestock feed grains could reduce trading opportunities and distort livestock sub-sectors within developing countries, with adverse consequences for poverty (Holden, 2002).

The impact of trade barriers is enormous. It has been estimated that developing country gains from a 50% cut in tariffs, by both developed and developing countries, would be in the order of US$150 billion (Nagarajan, 1999). Moreover, some of the non-tariff barriers, provisions and safety clauses in the World Trade Organisation agreement are being used to their advantage by various developed countries, resulting in serious trade distortions. The World Trade Organisation together with policy implications for livestock production in developing countries are discussed by Sharma (2002).

West African countries provide a good example of the implications of macro-economic, international trade and sectoral policies for the regional trade in livestock products (Williams *et al.*, 1995). The Sahel countries, over many centuries, were a major source of livestock products to their coastal neighbours. Major developments in the 1970s and 1980s changed the pattern of livestock trade in the region. The severe droughts of the late 1960s and early 1970s opened regional markets to frozen meat from South America and, later, the EU. This coincided with a sustained imbalance in macro-economic policies in the Sahel countries, and the emergence of considerable surpluses in the livestock sub-sector in the EU after years of heavy subsidies. As a result, internal beef exports to coastal West Africa dropped significantly. Imports of frozen beef, mainly from the EU, soared. The decline in Sahel exports occurred despite an almost complete recovery in the livestock sub-sector from the droughts of the 1970s. Although exports from the EU have now declined rapidly, the detrimental effects of EU trade policies may re-emerge if prices for Sahel beef increase and EU beef export subsidies also increase to facilitate the disposal of high stocks.

There seems to be a lack of agreement on the impacts of market liberalisation on the livelihoods of small-scale producers in developing countries, and strong reservations over the potential benefits. De Boer (1999) has concluded that, on balance, market liberalisation should help small-scale farmers in developing countries, as it will aid market expansion for dairy products and reduce the costs of importing inputs such as feed supplements and veterinary products. However, trade liberalisation will put increased pressure on high-cost producers of milk and milk products, and force smallholders to improve efficiency, cut costs and further rationalise collection, transport, processing and distribution. Evidence has been presented by Staal (2002) to indicate that market-orientated, smallholder dairy production will remain locally competitive in developing countries, in areas such as the tropical highlands of East Africa. He also presents evidence to suggest that, in some cases, countries with mainly smallholder dairy production systems will be able to compete internationally. In India, the dairy industry is again highly competitive, but must be protected from distorted and unfair trade competition in the liberalising economic environment (Sharma, 2002). The general opinion is that unless a level playing field is provided by the World Trade Organisation, the developing countries will always be in a disadvantaged position.

Conclusions

In the next 20 years, demand for meat and milk will more than double in developing countries, driven by changes in food preferences, and increases in human population, incomes and urbanisation. Market opportunities will provide an avenue for resource-poor farmers and small producers to improve their livelihoods. However, if projected targets are to be met, animal productivity will need to be increased significantly through the adoption of new technologies and the increased use of inputs. Technical and socio-economic constraints will have to be addressed within a favourable policy environment. Also, attention will need to be paid to the protection and better management of the environment. Threats to the improvement in the livelihoods of small farmers include negative environmental impacts in the production systems, involution, development of industrial systems of production, public health issues and inappropriate macro-economic policies.

The importance of mixed farming systems, as suppliers of ruminant animal products, is expected to grow in future along with non-ruminant industrial systems. Despite the threat of involution in some agro-ecological zones, the integration of crops and livestock will still represent the main avenue for the intensification of food production, and mixed farming will remain the predominant system in developing countries for some time to come.

References

Birthal, P.S. (1999). *Sources of growth in the livestock sector*. Policy Paper No.9. National Centre for Agricultural Economics and Policy Research, ICAR, New Delhi, India.

Birthal, P.S., Kumar, A. and Tewari, L. (ed.) (2000). *Livestock in different farming systems*. Proceedings of the annual conference of the Agricultural Economics Association, Tamil Nadu Veterinary and Animal Sciences University, Chennai, India.

Boer, A.J. de (1999). Socio-economic aspects of smallholder dairy farmers. In: *Smallholder dairying in the tropics*. Edited by Falvey, L. and Chantalakhana, C. International Livestock Research Institute (ILRI). Nairobi, Kenya, pp. 45-59.

Carter, J. (1995). Alley farming: Have resource-poor farmers benefited? *Natural Resources Perspectives* No. 3, 1-4.

Coker, R. (1999). Mycotoxins-the silent threat to human and animal health. An Inaugural Professorial Lecture delivered at the University of Greenwich, Chatham Maritime, UK, 24 March, 1999. University of Greenwich, London, UK.

Conroy, C. and Thakur, Y.A. (2002). Increasing the productivity of indigenous goat production systems through participatory research in ethno-veterinary medicine: A case study from India. In: *Responding to the increasing global demand for animal products - Programme and Summaries*. International conference organised by the British Society of Animal Science, American Society of Animal Science and Asociacion Mexicana de

Produccion Animal, held at the Universidad Autonoma de Yucatan, Merida, Mexico, 12-15 November, 2002. British Society of Animal Production, Penicuik, UK, pp. 67-68.

Delgado, C., Rosegrant, M., Steinfeld, H., Ehui, S., and Courbois, C. (1999). *Livestock to 2020: The next food revolution.* Food, Agriculture and Environment Discussion Paper 28, International Food Policy Research Institute (IFPRI), Washington, DC, USA.

Delgado, C.L., Rosegrant, M.W. and Meyer, S. (2001). *Livestock to 2020: The revolution continues.* http://www.agtrade.org/digests/digest33.htm.

Deoghare, P.R. and Sharma, B.M. (1992). Prospects of increasing farm income on marginal and small farms. *Agricultural Situation in India*, March issue, pp. 925-929.

Dev, S.M. (1998). Regional disparities in agricultural labour productivity and rural poverty. *Indian Economic Review* 23 (2): 167-205.

Devendra, C. (1995). Animal production systems in south-east and east Asia: Potential and challenges for research. In: *Global agenda for livestock research.* Edited by Gardiner, P. and Devendra, C. Proceedings of a Consultation held at ILRI, Nairobi, Kenya, 18-20 January, 1995. International Livestock Research Institute (ILRI), Nairobi, Kenya, pp. 49-57.

Devendra, C., Thomas, D., Jabbar, M.A. and Kudo, H. (1997). *Improvement of livestock production in crop-animal systems in rainfed agro-ecological zones of south-east Asia.* International Livestock Research Institute (ILRI), Nairobi, Kenya.

Devendra, C., Thomas, D., Jabbar, M.A. and Zerbini, E. (2000). *Improvement of livestock production in crop-animal systems in agro-ecological zones of south Asia.* International Livestock Research Institute (ILRI), Nairobi, Kenya.

Devendra, C. (2002). Crop-animal systems in Asia: Future perspectives. *Agricultural Systems* 71 (1-2):179-186.

Devendra, C. and Chantalakhana, C. (2002). Animals, poor people and food insecurity: Opportunities for improved livelihoods through efficient natural resource management. *Outlook on Agriculture* 31 (3):161-175.

Devendra, C. and Thomas, D. (2002). Smallholder farming systems in Asia. *Agricultural Systems* 71 (1-2):17-25.

Elbasha, E., Thornton, P.K. and Tarawali, G. (1999). An *ex post* economic assessment of planted forages in West Africa. ILRI Impact Assessment Series 2. International Livestock Research Institute (ILRI), Nairobi, Kenya.

FAO (1993). *FAO Agriculture: Towards 2010.* Food and Agriculture Organisation (FAO), Rome, Italy.

FAO (2002). *World agriculture: Towards 2015/30.* Summary Report. Food and Agriculture Organisation (FAO), Rome, Italy.

Gardiner, P. and Devendra, C. (ed.) (1995). *Global agenda for*

benefits and impacts on rural livelihoods. *Agricultural Systems* 71 (1-2):147-168.

Ramchandra, K.S., Anandan, S. and Raju, S.S. (2001). Livestock feed resources availability in India and the need for a national database. *Proceedings of the Xth Animal Nutrition Conference*, National Dairy Research Institute (NDRI), Karnal, India, pp. 134-144.

Rangnekar, D.V. (1988). *Integration of sugar and milk production in India*. FAO Animal Production and Health Paper No. 72. Food and Agriculture Organisation (FAO), Rome, Italy.

Rangnekar, D.V. (1995). Research methodology for crop-animal systems in semi-arid regions of India. In: *Crop-animal interaction*. Proceedings of an international workshop held at Khon Kaen, Thailand, 27 September-1 October 1993. International Rice Research Institute (IRRI), Manila, Philippines, pp. 405-420.

Rangnekar, S. D. (1998). The role of women in smallholder rainfed and mixed farming in India. In: *Women in agriculture and modern communication*. Proceedings of a workshop, Tune, Denmark, 30 March-13 April 1998, pp. 25-37.

Raja, R.H. (2002). Pakistan smallholder dairy production and marketing. In: *Smallholder dairy production and marketing - opportunities and constraints*. Edited by Rangnekar, D. and Thorpe, W. Proceedings of a South-South workshop held at NDDB, Anand, India, 13-16 March, 2001. National Dairy Development Board (NDDB), Anand, India and International Livestock Research Institute (ILRI), Nairobi, Kenya, pp. 47-64.

Renard, C. (ed.) (1997). *Crop residues in sustainable mixed crop/ livestock farming systems*. CAB International, Wallingford, UK.

Richards, J.I. (2001). Introduction to the DFID Livestock Research Programme. In: *Sustaining livestock in challenging dry season environments: strategies for smallscale livestock farmers*. Edited by Smith, T. and Godfrey, S.H. Proceedings of the Third Workshop on Livestock Production Programme projects held at Ingwe Lodge and ICRISAT, Matobo, Zimbabwe, 26-28 September, 2000. Natural Resources International Ltd, Chatham Maritime, UK, pp. 11-12.

Riveros, F., Crespo, D. and Ben Ali, M.N. (1993). Constraints to introducing the ley farming system in the Mediterranean basin. In: *Introducing ley farming to the Mediterranean basin*. Edited by Christiansen, S., Materon, L., Falcinelli, M. and Cocks, P. International Centre for Agricultural Research in Dry Areas (ICARDA), Aleppo, Syria, pp. 15-22.

Ruiz, M. (1995). The Latin American livestock sector and research prospects. In: *Global agenda for livestock research*. Edited by Gardiner, P. and Devendra, C. Proceedings of a Consultation held at ILRI, Nairobi, Kenya, 18-20 January, 1995. International Livestock Research Institute (ILRI), Nairobi, Kenya, pp. 61-73.

Sampark (1997). Impact study of milk cooperatives at village level-Karnataka study. In: *OED-IRMA workshop on the impact of Operation Flood*. Institute of Rural Management, Anand, India. 17-18 March 1997.

Sansoucy, R., Jabbar, M.A., Ehui, S. and Fitzhugh, H. (1995). The contribution of livestock to food security and sustainable development. In: *Livestock strategies for low income countries.* Edited by Wilson, R.T., Ehui, S. and Mack, S. Proceedings of the Joint FAO/ILRI Roundtable held at ILRI, Addis Ababa, Ethiopia, 27 February-2 March, 1995. Food and Agriculture Organisation (FAO) and International Livestock Research Institute (ILRI), Nairobi, Kenya, pp. 9-21.

Schelhaas, H. (1999). The dairy industry in a changing world. In: *Smallholder dairying in the tropics*. Edited by Falvey, L. and Chantalakhana, C. International Livestock Research Institute (ILRI), Nairobi, Kenya, pp. 1-17.

Sere, C. and Steinfeld, H. (1996). *World livestock production systems: current status, issues and trends*. FAO Animal Production and Health Paper 127. Food and Agriculture Organisation (FAO), Rome, Italy.

Sharma, V.P. (2002). Implications of international trade regulations (World Trade Organisation Agreement on Agriculture and Codex Standards) for smallholder dairy development. In: *Smallholder dairy production and marketing - opportunities and constraints*. Edited by Rangnekar, D. and Thorpe, W. Proceedings of a South-South workshop held at NDDB, Anand, India, 13-16 March, 2001. National Dairy Development Board (NDDB), Anand, India and International Livestock Research Institute (ILRI), Nairobi, Kenya, pp. 231-249.

Shukla, R.K., Dass, B., Singh, B. and Yadav, S.R. (1995). Impact of Operation Flood Programme on the economy of rural milk producers in District Kanpur, Uttar Pradesh. ICAER, New Delhi, India.

Shukla, R.K. and Brahmankar, S.D. (1999). *Impact evaluation of Operation Flood on rural dairy sector*. ICAER, New Delhi, India.

Singh, R. (1995). Role of dairying as a specialised enterprise in poverty alleviation of the landless and marginal farmers of rural Azamgadh in Uttar *Pradesh. Indian Journal of Agricultural Economics* 50(3):352 (abstr.).

Singh, K., Habib, G., Siddiqui, M.M. and Ibrahim, M.N.M. (1997). Dynamics of feed resources in mixed farming systems of South Asia. In: *Crop residues in sustainable mixed crop/livestock farming systems*. Edited by Renard, C. CAB International, Wallingford, UK, pp. 113-130.

Singh, C.B., Agarwal, S.B., Chandra, N., Jha, S.K. and Dey, R.A. (1998). Crossbreeding technology adoption and constraint identification in different regions of India. Report. National Dairy

Research Institute, Karnal, India.

Smith, T., Moyo, S., Richards, J.I.and Morton, J.F. (2002). The role of indigenous and crossbred cattle for smallholder dairy production in Zimbabwe. In: *Responding to the increasing global demand for animal products - Programme and Summaries*. International conference organised by the British Society of Animal Science, American Society of Animal Science and Asociacion Mexicana de Produccion Animal, held at the Universidad Autonoma de Yucatan, Merida, Mexico, 12-15 November, 2002. British Society of Animal Production, Penicuik, UK, pp. 223-224.

Staal, S.J. (2002). The competitiveness of smallholder dairy production: Evidence from sub-Saharan Africa, Asia and Latin America. In: *Smallholder dairy production and marketing - opportunities and constraints*. Edited by Rangnekar, D. and Thorpe, W. Proceedings of a South-South workshop held at NDDB, Anand, India, 13-16 March, 2001. National Dairy Development Board (NDDB), Anand, India and International Livestock Research Institute (ILRI), Nairobi, Kenya, pp. 250-264.

Steinfeld, H., Haan, C. de, and Blackburn, H. (1997). *Livestock-environment interactions: issues and options*. FAO/USAID/World Bank, WRENmedia, Eye, Suffolk, UK.

Steinfeld, H. (2001). Livestock production towards 2020. In: *Proceedings of the first interagency meeting on livestock production and animal health*. Edited by Hainsworth, S.D., Godfrey, S.H., Matthewman, R.W. and Richards, J.I., Meeting held at Imperial College at Wye, Wye, UK, 6-7 December 2000. Natural Resources International Ltd, Chatham Maritime, UK, pp. 7-13.

Thomas, D. and Sumberg, J.E. (1995). A review of the evaluation and use of tropical forage legumes in sub-Saharan Africa. *Agriculture, Ecosystems and Environment* 54:151-163.

Thomas, D. (1999). Ruminants in crop-animal systems in developing countries: beasts or benefactors? An Inaugural Professorial Lecture delivered at the University of Greenwich, Chatham Maritime, United Kingdom, 10 February, 1999. University of Greenwich, London, UK.

Thomas, D., Zerbini, E., Parthasarathy Rao, P. and Vaidyanathan, A. (2002). Increasing animal productivity on small mixed farms in South Asia: A systems perspective. *Agricultural Systems* 71 (1-2): 41-57.

Thornton, P.K., Kruska, R.L., Henninger, N., Kristjanson, P.M., Reid, R.S., Atieno, F., Odero, A. and Ndegwa, T. (2001). *Mapping poverty and livestock*. Report to DFID. International Livestock Research Institute (ILRI), Nairobi, Kenya.

Vercoe, J., Coffey, S., Farrell, D.J., Rutherford, A. and Winter, W.H. (1997). *ILRI in Asia: an assessment of priorities for Asian livestock research and development*. International Livestock Research

Institute (ILRI), Nairobi, Kenya.

White, J. and Robinson, E. (2000). *HIV/AIDS and rural livelihoods in Sub-Saharan Africa*. Policy Series 6, Centre for Sustainable Development. Natural Resources Institute, Chatham Maritime, UK.

Williams, T.O., DeRosa, D.A. and Badiane, O. (1995). Macroeconomic, international trade and sectoral policies in livestock development. In: *Livestock strategies for low income countries*. Edited by Wilson, R.T., Ehui, S. and Mack, S. Proceedings of the Joint FAO/ILRI Roundtable held at ILRI, Addis Ababa, Ethiopia, 27 February-2 March 1995. Food and Agriculture Organisation (FAO) and International Livestock Research Institute (ILRI), Nairobi, Kenya, pp. 47-69.

Winrock (1992). *Animal agriculture in Sub-Saharan Africa*. Winrock International, Morrilton, USA.

World Bank (1996). *India livestock sector review: enhancing growth and development*. Report No. 14522-IN. Agriculture and Water Operations Division, Country Department 11, South Asia Region. World Bank, Washington, DC, USA.

World Bank (2000). *World development report 2000*. World Bank, Washington DC, USA.

2

Mapping livestock and poverty: a tool for targeting research and development

P.K. Thornton[1,2], P.M. Kristjanson[1], R.L. Kruska[1] and R.S. Reid[1]
[1]*International Livestock Research Institute (ILRI), PO Box 30709, Nairobi 00100, Kenya*
[2]*Institute of Atmospheric and Environmental Sciences, University of Edinburgh, Edinburgh EH9 3JG, UK*

Introduction

Detailed geographical profiles of poverty and inequality can be extremely valuable to governments, non-governmental organizations, and multilateral institutions that aim to strengthen the impact of their spending on poverty. Some developing countries use information on the geographical distribution of welfare indicators to guide the allocation of resources among local agencies or government as a first step in reaching the poor. Poverty maps can also be important tools for researchers. In addition to visualizing the incidence, magnitude and heterogeneity of poverty through space, georeferenced welfare estimates can be used to examine poverty linkages with other spatially explicit socio-economic and bio-physical variables. Given many organisations' focus on poverty reduction, there is an urgent need for research and development agencies to reconsider how best to operate in ways that will benefit poor people. However, our knowledge of the location, characteristics and trends of change among poor livestock keeping populations is very patchy, both spatially and temporally.

In this paper, we outline a rapid broad-brush global assessment of spatial distribution of poor livestock keepers, and describe parallel activities in high-resolution poverty mapping for countries in East Africa using sophisticated econometric techniques pioneered at the World Bank. We then describe some of the ways in which this information is being utilised, in conjunction with other activities such as intervention mapping, livelihoods mapping, and survey work, to improve our understanding of the situations where livestock can offer the means for helping to alleviate poverty, and the mechanisms whereby this can be brought about.

Locating poor livestock keepers globally

In 2001 the UK Government's Department for International

Table 2.2. Number of "poor livestock keepers" (millions) by region and livestock production system: nationally-defined rural poverty rates (World Bank, 2001; LID, 1999).

Region	*Rangland-based systems*	*Irrigated mixed Systems*	*Rainfed mixed Systems*	*Landless systems*	*Total*
East Asia	1.3	11.3	15.2	0.2	28.0
Central and South America	10.4	4.6	71.2	5.1	91.3
Central Asia	2.1	1.9	5.2	<<0.1	9.2
South Asia	6.3	71.8	127.0	0.7	205.9
South-East Asia	0.3	17.2	43.6	0.7	61.7
Sub-Saharan Africa	20.5	0.9	139.8	0.7	162.0
West Asia-North Africa	5.6	7.1	26.4	0.2	39.3
Total	46.3	114.8	428.5	7.6	597.3

million poor livestock keepers, and the mixed rainfed systems some 430 million poor livestock keepers. In terms of the magnitude of poverty and the importance of livestock to poor households in the developing world, this analysis suggests that there are nearly 600 million poor livestock keepers globally.

Second, population growth and climate change will produce substantial changes in livestock production systems over the next three to five decades. There are indications that the magnitude of these systems changes, and the consequent need for adaptation and mitigation work, will be particularly large in sub-Saharan Africa; these analyses remain to be completed for Asia and Latin America.

Third, considerably more work is required to better inform donors and the research and development community of where hotspots of change are located, who is likely to be affected, and how. Global-level analyses can effectively (if only broadly) identify foci where research and development activities aimed at specific groups of people might profitably be targeted. However, there are also substantial needs for within-in country poverty mapping activities, to allow for much higher-resolution targeting. Below, we outline some of these, and show the types of analysis that are possible with the resulting information.

High-resolution poverty maps in Kenya

A research collaboration between ILRI, Kenya's Central Bureau of Statistics, Rockefeller and World Bank researchers has resulted in the

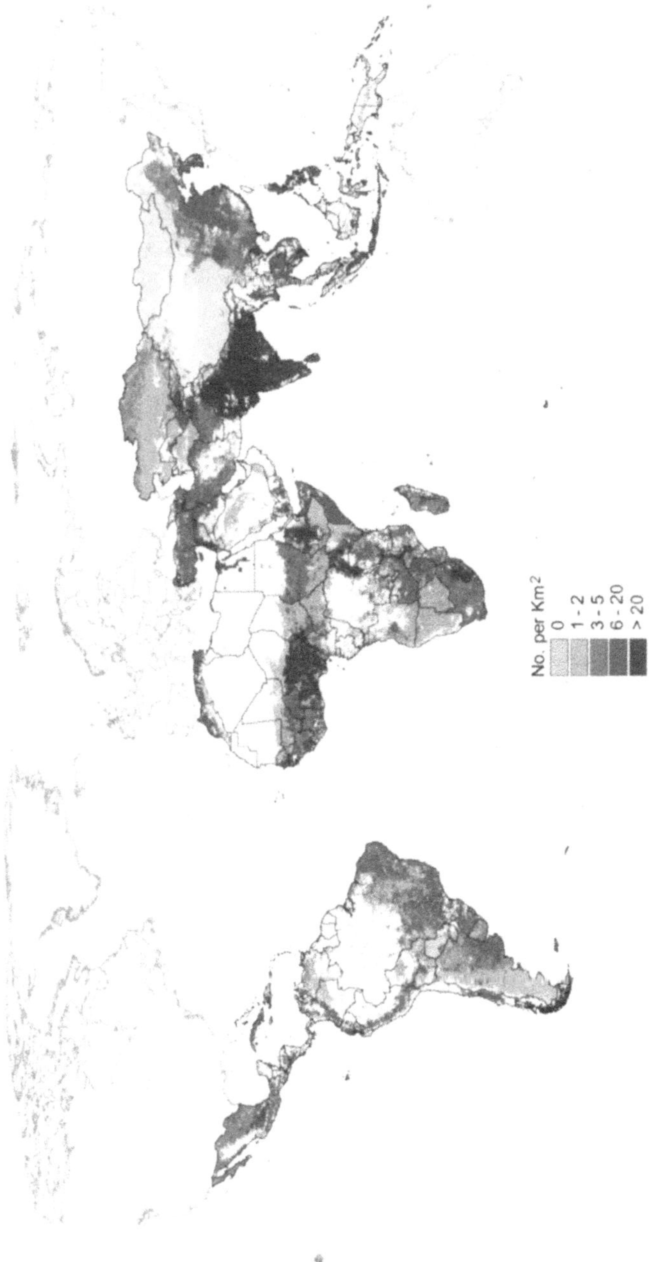

Figure 2.1. Density of "Poor livestock keepers" (Thornton *et al.*, 2002, 2003).

first high-resolution poverty maps for Kenya (maps for Uganda will also soon be available). Estimates of the percentage of the population living below the poverty line at the location-level (thousands of households) are the result of this teamwork (CBS, 2003). This research is now entering the dissemination and use stage, with exciting possibilities for using the results directly for policy targeting and better understanding spatial patterns of poverty in Kenya, but also for the further research opportunities that have been opened up to analyze the underlying determinants of these patterns.

The technique used for the Kenya poverty map is based on small area estimation (Hentschel and Lanjouw, 1996; Hentschel *et al.*, 1998; 2000; Elbers *et al.*, 2002; 2003). Essentially, the approach involves taking a national census and a national representative household survey, such as a Welfare Monitoring Survey (WMS). A set of variables common to both the census and the survey is identified, and a model is built to estimate household expenditure from the survey. This model is then applied to the census data at household unit level to predict *per caput* household expenditure. The household data can then be aggregated to obtain robust estimates of the percentage of households living below the poverty line, which can then be mapped spatially. In most cases, the spatial resolution of the map is significantly higher than would be possible with household survey alone (Henninger and Snel, 2002). This approach has been undertaken in various countries around the world, and is currently underway for six countries in East and southern Africa. The activities in Kenya are allowing us to address complex issues relating to the links between land-use, land degradation, and poverty. The high-resolution maps that are being produced will help provide much stronger links between information and decision-making at local levels, a critical step towards reduction of poverty in the developing world.

One of the research opportunities afforded by the high-resolution poverty maps involves spatial analysis to examine broader welfare issues (Demombynes *et al.*, 2002; Mistiaen *et al.*, 2002) and poverty-environment linkages, in the attempt to improve poverty reduction strategies and targeting. Because locations and sublocations are relatively small areas compared with districts, spatial factors such as soil, climate and even market access among households within a location or sublocation are relatively homogenous. Such high-resolution data for a welfare measure will allow an investigation of the links between geographic characteristics and poverty. This is an analysis that needs to be done at a scale that provides enough variation in critical spatial variables (such as temperature, altitude, access to markets, ethnicity, etc) to examine the importance of these factors in determining land-use patterns and relative poverty levels. If such "map-able" geographic and socio-economic factors are in fact largely determining the degree of expenditure or welfare levels, this might not show up in the results of typical household-level studies that are confined to a small area, simply because of lack of variation. At the same time, household-level census or survey data that can be aggregated to the same broader level of observation (e.g. location) can be used to capture "traditional" determinants of poverty, such as level of education. The premise behind this spatial analytical approach, therefore, is that apart from traditional household-level factors, spatial factors are likely to be crucial in understanding levels of poverty at the broader landscape or community level. Using the current GIS database developed at ILRI, it should be possible to evaluate agro-climatic characteristics such as amount, distribution and variability of rainfall, soil type and slope, to allow a more in-depth examination of the linkages between poverty and environmental degradation than was previously possible. Such analyses are now possible because high-resolution poverty data and new remote sensing data on land cover are now available. This research will also build on previous ILRI analysis that shows that market access is of considerable importance in household decision making related to dairying (Staal *et al.*, 2002). ILRI has developed a highly detailed road network database for the Kenyan Highlands that has been successfully used in the past to characterise market access (see Figure 2.2), and this is currently being extended in order to cover the whole country.

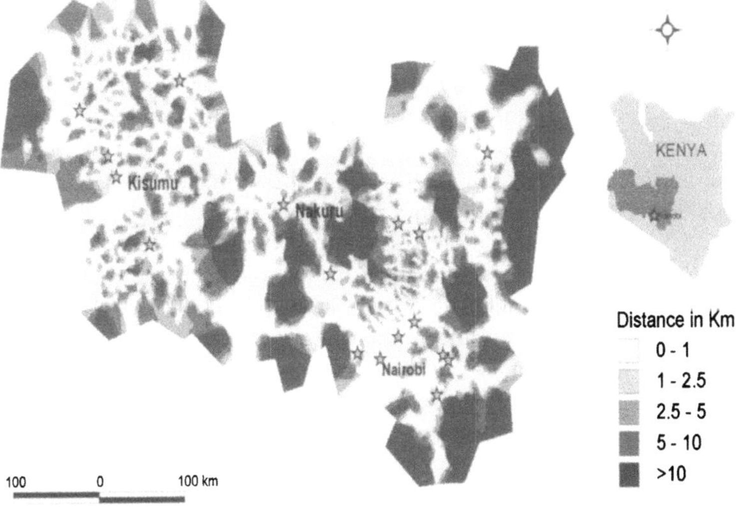

Figure 2.2. Distance to the nearest good-quality road: Kenya Highlands (source: ILRI databases).

The analysis is combining data from different sources, evaluated at the lowest or second lowest administrative level in Kenya (sublocation or location). The variable of interest (main dependent variable) is the level of poverty, defined as the percentage of households falling below a specific poverty line. Two major sets of explanatory variables are identified: socio-economic factors available through census data, and spatial factors derived from existing GIS layers, both purely biophysical (such as altitude and soil type) as well as some with socio-economic aspects (such as market access). The combined use of these rich datasets should allow us to disentangle the influence of environmental conditions and socio-economic characteristics on the level of poverty and will lead to important policy recommendations as well as information that can be used to improve targeting of research and development (and policy) interventions. For example, controlling for agro-climatic characteristics, if socio-economic factors such as market access turn out to significantly influence the level and distribution of poverty, strong policy recommendations in terms of road infrastructure and/or marketing channels can be derived.

Poverty intervention mapping and development of new tools

The new poverty maps also open up opportunities for developing tools aimed at improved targeting of research investments in terms of interventions that can reach the poorest segments of society. A good understanding of where particular livestock-related interventions (technologies, strategies, policies) are feasible and relevant is needed. This can be done by developing an inventory of interventions and characterizing them. Once this is done, areas can be mapped that match the range of biophysical and socio-economic factors relevant to (or required by) particular interventions, such as the amount of rainfall and the proximity to markets. Subsequent analysis can then be used to estimate the size and approximate locations of potential target groups for particular interventions.

A simple example is shown in Figure 2.3, for Kajiado District, Kenya. Despite that fact that the predominant livestock system in Kajiado District is pastoralism, livestock keepers are increasingly planting crops to supplement household food demands. Millet and sorghum suitability zones are mapped in Figure 2.3 according to rainfall and average temperature: 400-600 mm rainfall and 20.1 to 22.2 °C for millet, and 520-720 mm rainfall and 18.0 to 20.0 °C for sorghum. When overlaid with location-level poverty rates, there would appear to be some association between sorghum and millet suitability and lower poverty rates. This is not surprising, in that households located in agro-ecological zones that are more conducive to cropping might be expected to have more options for food security and income generation than others. Figure 2.3 can then be overlaid with current and future human population densities. For 2000, there are some 133,000 people in the millet zone (as mapped), and 72,000 people in the sorghum zone, of a total population for Kajiado District of 754,000. Using the projections of Reid *et al.* (2000), these numbers

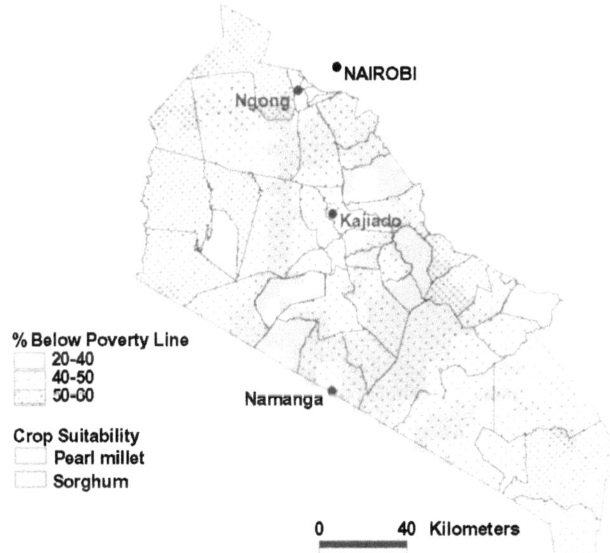

Figure 2.3. Kajiado District, Kenya, and areas suitable for growing pearl millet and sorghum overlaid with location-level poverty rates.

increase to 226,000 for the millet, 125,000 for the sorghum, and 1.28 million for Kajiado as a whole, to 2050. The implications on household food security and on land-use change of sustaining such numbers of people pose a number of difficult questions.

While maps such as these are useful for some users, even more useful for others will be a web- and CD-based "policy intervention" tool that will allow the user to define the relevant ranges and develop their own domains and policy-related queries. This is the goal of an ongoing ILRI activity, with funding from the UK Department for International Development. A prototype of the tool, tentatively named PRIMAS (Poverty Reduction Intervention Mapping in Agricultural Systems), is expected to be available for testing by mid-2004.

Livelihood maps

Recognizing that expenditure-based poverty estimates do not necessarily capture the importance of all the five types of assets (physical, natural, social, financial and human – see Table 2.3) that may largely determine the livelihood options facing poor households, ILRI is collaborating with various partners, including the Food Insecurity and Vulnerability Mapping Systems (FIVIMS) programme of FAO, on a livelihoods mapping project, commencing with Kajiado District in Kenya. This project is exploring what is "map-able" with respect to these five types of assets. Part of this work revolves around capturing the idea of what are the relevant catchment areas for assets such as schools, health facilities, agricultural services (crops and livestock),

Table 2.3. Examples of the five types of livelihoods assets that can be mapped.

Asset type	Map-able examples
Physical	Markets, roads, water, electricity, telephone services, livestock dips
Natural	Agricultural potential: rainfall (amount, distribution, variability), soils, water sources, communal forage land, forests
Human	Agricultural services, including skilled people such as vets, other service providers, health facilities/services, human disease incidence and sickness rates, availability of labour
Social	Information networks and groups, churches, community-based organizations, NGOs
Financial	Credit, livestock, off-farm income options (such as mining sites, tourism)

access to water, and communal land for grazing. These might be related to the travel times required to reach these facilities and services. This will allow us to identify areas that are not serviced, under-serviced, or sufficiently serviced with respect to a variety of facilities and services. We are also looking at the relationship between different types of services (do they all clump together, and are they mostly found only in large towns?). Another objective is to identify areas that have high and low levels of physical and natural assets, and compare areas of high and low levels of poverty (using the high-resolution poverty maps).

Community and household-level surveys

The approaches above can give us a better understanding of the some of the geographical determinants of poverty as well as some of the underlying factors determining households' livelihood options. None of them really gets at the underlying processes of poverty, however. ILRI has just completed a pilot test study in western Kenya that was aimed at testing a method for better understanding pathways out of and into poverty and the role livestock play (Kristjanson *et al.*, 2003). The method involves participatory community-level discussions using a facilitative approach, followed by household-level discussions. The use of a simple diagnostic tool makes this approach analytical, yet thoroughly participatory in nature. The results of the pilot test were interesting and encouraging, and the plan is to broaden out the study to cover 40 villages in Kenya and 40 in Uganda. Villages are being chosen using GIS techniques to stratify them according to several criteria that are important in defining the range of livelihood options available to villagers, including degree of market access and agricultural potential.

This project will generate knowledge as to what major factors are responsible for households' declines into poverty and the various reasons that other households have managed to boost themselves out of poverty (as defined by the communities themselves, rather than a government-defined poverty line that means little to them). These reasons are likely to vary significantly in different types of villages in different regions. For example, human disease and funeral-related costs are likely to be a more important factor explaining declines into poverty in drier areas near Lake Victoria, as was seen in the pilot project, whereas livestock ownership is likely to be a more critical factor in the higher-rainfall and higher-altitude areas of western Kenya. This information will allow the development at the regional or national level of sector-specific investments and policies for particular areas and certain types of communities. For instance, if the most important reason within a particular district for declines into poverty is the non-availability of consumer credit, a possible solution lies in working with communities to develop an appropriate micro-credit scheme. If, however, in another district, it is found to be a particular disease, a community-driven disease eradication program may do more to prevent needless declines into poverty than will a credit scheme. More contextualized and precise knowledge about poverty will be generated, and better targeted programs can be formulated to utilize scarce development resources more effectively.

Conclusions

The notion of poverty has many dimensions, and there are considerable complexities simply in attempting to spatially locate where poor people are, let alone trying to understand the reasons for poverty existing in certain places. It is clear from the above discussion that there are different scales and different tools that have to be used to improve our understanding of poverty and its relation to rural populations and livestock.

All the work above is designed to answer questions relating to improved targeting of interventions and improved understanding of the role of livestock in poverty alleviation. The global mapping study has been used within ILRI to assess a wide range of animal health research and development opportunities in the tropics (Perry *et al.*, 2002), and is also being used within DFID to assist in targeting and setting priorities (S. Holden, personal communication). The higher-resolution work, in time, should greatly improve our understanding of the proportion of income that people in different production systems derive from livestock, and thus the importance of livestock to their livelihoods, and how this is likely to change in the future as a result of factors such as population increase, climate change, and systems evolution.

Ecosystems and Environment 77: 227-236.

Seré, C. and Steinfeld, H. (1996). *World livestock production systems: current status, issues and trends*. FAO Animal Production and Health Paper 127. FAO, Rome.

Staal, S.J., Baltenweck, I., Waithaka, M.M., de Wolff, T. and Njoroge, L. (2002). Location and uptake: integrated household and GIS analysis of technology adoption and land use, with application to smallholder dairy farms in Kenya. *Agricultural Economics* 27 (3): 295-315.

Thornton, P.K., Kruska, R.L., Henninger, N., Kristjanson, P.M., Reid, R.S., Atieno, F., Odero, A. and Ndegwa, T. (2002). *Mapping poverty and livestock in the developing world*. ILRI, Nairobi, Kenya.

Thornton, P.K., Kruska, R.L., Henninger, N., Kristjanson, P.M. and Reid, R.S. (2003). Livestock and poverty maps for research and development targeting in the developing world. *Land Use Policy* 20: 311-322.

United Nations Development Programme (UNDP) (1997). *Human development report 1997*. Oxford University Press, Oxford, UK.

World Bank (2001). *World development report 2000/2001: Attacking poverty*. Oxford University Press, New York.

3

The Livestock Revolution - a view on implications for Africa

D.M. Mwangi and A. Omore
Kenya Agricultural Research Institute, PO Box 30148, Nairobi, Kenya

Introduction

A rapid growth in *per capita* consumption of livestock products is occurring in developing countries (Delgado *et al.*, 1999; FOASTAT 1983-2002). In the last two decades and until 2020, the increase in the consumption of meat and milk has been estimated to be much higher in developing countries compared to developed countries (Delgado *et al.* 1999). This increase in consumption and demand is fuelling an increase in production.

Developing countries, that only produced 51 M MT (million tonnes) or 36% of the world's meat in 1983, recently surpassed the developed world in production, an estimated 132 M MT, or 55% of the world's total meat production. This share is predicted to increase further to 183 M MT, or 60% by 2020 (Table 3.1). Corresponding increases and share of milk production reflect similar treads: with a consumption of only 113 M MT or 24% of the world's milk in 1983, the quantity produced has more than doubled to 245 M MT or 41% of world production currently, and is further predicted to reach 400 M MT or 52% by 2020. This on-going increase in production of livestock products by developing compared to developed countries, or the "livestock revolution", is mainly in response to large increases in demand driven by human population growth, urbanization and modest increases in incomes (Delgado *et al.*, 1999). Though the changes represent important opportunities for transformation of livestock marketing and production in developing countries, major obstacles need to be overcome to allow the currently-dominant small-scale producers and market agents to benefit from the revolution.

The livestock revolution and challenges for Sub-Saharan Africa (SSA)

Though the large increases in consumption and production of livestock

Table 3.1. Past and projected trends in consumption and production of livestock products.

	Sub-Saharan Africa			Developing world			Developed world		
	1983	1993	2020	1983	1993	2020	1983	1993	2020
Consumption									
Annual *per capita* meat consumption (kg)	10	9	11	14	21	30	74	76	83
Annual *per capita* milk consumption (kg)	32	23	30	35	40	62	195	192	189
Total meat consumption (million MT)	4	5	12	50	88	188	88	97	115
Total milk consumption (million MT)	12	14	31	122	168	391	233	245	263
Production									
Annual *per capita* meat production (kg)	10	9	10	15	21	29	76	78	87
Annual *per capita* milk production (kg)	19	19	30	32	39	63	305	272	267
Total meat production (million MT)	-	-	11	51	88	183	90	100	121
Total milk production (million MT)	-	-	31	113	164	401	365	348	371
Shares of meat production (%)	3	3	-	36	47	-	64	53	-
Shares of milk production (%)	2	2	-	24	32	-	76	68	-

Source: Delgado *et al.* (1999)

products from 1983-94, occurred in developing countries, the increase was lowest in SSA. Since then the continent has caught up (Table 3.2). The SSA share of both meat and dairy production from developing countries that was 8-9% in 1983, marginally declined to 5-7% currently and is predicted to remain the same in the coming two decades (including North Africa would increase the proportions by some 60-80% but the trends remain the same; henceforth the focus will be on SSA). Most of this growth is occurring from increases in herd size, rather than in productivity (Table 3.3). Up to now, *per capita* production of livestock products has kept up with demand, except in SSA for meat and West Asia and North Africa for milk (Dalgado *et al.* 1999). Though *per capita* production in SSA is still relatively low at only one-third for meat and one-half for milk compared to the average for developing countries, the on-going increases in demand definitely offer new opportunities for agricultural growth.

For SSA to meet the increasing demand for livestock products from domestic and regional production (and reduce the need to import these products thereby benefiting local producers who are mainly poor livestock keepers), several constraints to production must be addressed. The current contribution of livestock production to African economies indicates widespread under-utilization and low productivity, as exemplified by Tanzania, which has the third largest livestock herd in Africa, after Ethiopia and Sudan. Tanzania has an estimated 15.6 million cattle, 10.7million goats and 3.5 million sheep kept by 37 per cent of the 3.87 million agricultural households (Bureau of Statistics, 1996). Despite this large herd and the fact that more than three quarters on the country is only suitable for livestock production, the sector only contributes 18% to the GDP (Bureau of Statistics, 1996). Similarly in Kenya, the livestock sector contributes only 10% to the GDP yet uses 84% of the land. However, the land is arid and semi-arid, and suitable only for livestock production (Anon, 2001). This same scenario is repeated in several other African countries. This low contribution of the livestock sector to the economy of most African countries can be attributed to the quality of breeds currently kept, low quality feeds, diseases, lack of, or inadequate, support services to livestock keepers, inappropriate policies and low access to input and output markets.

Required interventions in improved livestock genetics

One avenue for responding to the projected increases in demand for livestock products in SSA is adoption of livestock with genetics for improved production. Currently most livestock production systems area based on low-producing indigenous animals due to their resistance/tolerance to prevailing diseases and capacity to utilize low

Table 3.2. Projected trends in growth in consumption and production of livestock products.

%	Sub-Saharan Africa		Developing world		Developed world	
	1982-1994	*1993-2020*	*1982-1994*	*1993-2020*	*1982-1994*	*1993-2020*
Annual growth rate of meat consumption :	2.2	3.5	5.4	2.8	1.0	0.6
Annual growth rate of milk consumption :	-	3.8	3.1	3.3	0.5	0.2
Annual growth rate of meat production :	2.1	3.4	5.4	2.7	1.1	0.7
Annual growth rate of milk production :	-	4.0	3.7	3.2	-0.4	0.4

Source: Delgado *et al.* (1999)

Table 3.3. Projected trends in growth in productivity and productivity growth rate.

	Sub-Saharan Africa		Developing world		Developed world	
	Productivity kg/head	*Productivity growth rate %/yr*	*Productivity kg/head*	*Productivity growth rate %/yr*	*Productivity kg/head*	*Productivity growth rate %/yr*
	1992-94	1982-94	1992-94	1982-94	1992-94	1982-94
Beef	152	-0.5	162	0.5	242	0.9
Milk	340	0.6	896	1.9	3,739	1.3
Pork	45	0.1	72	1.2	82	0.4
Poultry	0.9	-0.1	1.2	0.8	1.3	0.7

Source: Delgado *et al.* (1999)

quality feeds. The Kenya dairy industry is a good example where upgrading of the local breeds using exotic dairy genes, especially the Friesian and Aryshire, and change in production system have transformed the livelihoods of many. Before independence African farmers were not allowed to keep exotic dairy breeds and the dairy sector in the country was dominated by large scale dairy farms owned by European settlers. At independence in 1963, the new government in Kenya encouraged African farmers to upgrade their indigenous animals by subsidising artificial insemination services. The upgrading was accompanied by intensive use of the land for the production of pasture. As the land available for pasture production declined due to land sub-divisions necessitated by human population growth, the cut-

and-carry production system developed, with high yielding forage crops including Napier grass (*Pennisetum purpureum*) replacing pasture (Staal *et al.*, 2002). This process has transformed the dairy sub-sector in Kenya from one dominated by large scale farmers to one where over 600,000 smallholder dairy farms, consisting of mainly family farms, typically with an average of 2-3 animals on a hectare or less land–_produce approximately 80% of the estimated 1.092 M MT milk marketed in Kenya annually (Omore *et al.*, 1999). Similar benefits have been realised through use of improved livestock genetics in other places. For example in the Kodowa area in Tanzania, household income and food security have improved with adoption of crosses between local goats and exotic dairy breeds and a change of production system. Unlike in the Kenyan case, the change in the production system in Kodowa was mainly aimed at guarding against soil erosion in the already seriously eroded area (Shirima and Msechu, 2003).

Required interventions in improved nutrition

Apart from improving the genetic potential of livestock, responding to and deriving benefits from the livestock revolution will also require investments in attaining higher feed availability, especially during the dry season. Most of the livestock production systems in Africa are characterized by low and cyclic growth rates, high calf mortality and long calving intervals resulting in reproductive wastage. In the Kenya dairy sector calving intervals are as long as 600 days (Odima *et al.*, 1994), calf growth rates are well below 500 g/day and calf mortality stands at 20% (Gitau *et al.*, 1994). This is mainly attributed to under nutrition and poor management of the animals. Therefore, to meet the increased demand for livestock products in Africa, the issue of improved management and matching breeds to available feed resources must be addressed. An option for addressing under nutrition is to enhance the adoption of technologies which will improve the utilization of cereal residues available in the mixed farming systems that dominate in arable areas.

In the rangelands, the predominant source of livelihoods is ruminant production, which is wholly dependent on natural grasslands. Extreme weather conditions such as droughts have large adverse effects on feed resources and production. Ndikumana *et al.* (2000), who have recently studied the impacts of drought and the coping mechanisms employed by pastoralists, report that mortality rates rise to over 50% in extreme droughts. Though rainy weather is beneficial to forage production, the potential benefits of abundant feed is often reduced by high morbidity and mortality rates from vector-borne diseases. Appropriate measures, such as early warning systems for adverse weather conditions, will need to be put in place to mitigate such losses.

al., 1999). While smallholders and informal raw milk market channels dominate and satisfy the supply of marketed dairy products in East Africa (up to 98% of market milk sales in Tanzania), in Southern Africa (mainly Zimbabwe and South Africa) supply from large-scale dairy farms and cold-chain market channels predominate and also satisfy local demand. In West Africa, marketed domestic dairy production is limited, necessitating large quantities of dairy imports. Kenya has the highest *per capita* availability of 90 LME (litres of milk equivalent) which is four to seven times higher than the other countries in SSA. The high level of marketed production in Kenya is mostly supplied by its 3 million crossbred and high-grade dairy cattle in smallholdings that comprise 70% of all such cattle in eastern and southern Africa. Marketed domestic dairy supply from crossbred or high-grade cattle in other central and southern African countries is very limited. In well-populated areas of Nigeria and Ghana, milk supply is dominated by Fulani agro-pastoralists, while in Mali's capital, Bamako, there are high-input large-scale dairies using crossbred and high grade cattle close to urban areas. Their dairy production is supplemented by peri-urban communal dairies producing milk from herds of local cattle assembled from several owners in peri-urban areas. In these locations, the dairy production systems and their market supply channels will probably remain based on traditional methods in the foreseeable future, but are likely to change as demand and incomes increase resulting in shifts in tastes and preferences in favour of processed dairy products. In the liberalized market and the improving economy of Dar-es-Salaam some of these changes are already being observed.

Innovation is a major driving factor in development. If Africa has to overcome the constraints stated above to enable small-scale livestock keepers to gain from the livestock revolution, then African governments will have to invest more in science and technology. Currently African countries, with the exception of South Africa, are investing less than 1% of GDP in science and technology. In the case of Kenya, only 0.01% of GDP is invested in science and technology. Therefore, most of the funds invested in science and technology comes from donor funded projects.

The livestock revolution and the poor

Though livestock help poor farmers in Africa to accumulate capital and to have an inflation free form of banking, the start-up capital, especially for improved breeds is often too high for poor farmers. In central Kenya an in-calf dairy heifer will cost between KSh. 30,000 – KSh. 100,000 (US$ 429 – 1,429) and produce milk of roughly equivalent value in the first lactation. Though returns are realised quickly, in an area where household monthly income is about KSh

5,831 (US$. 83.3), the capital cost of purchasing a heifer is out of reach for most poor farmers. Availability of affordable credit is, therefore, an important intervention. Currently, most poor farmers make do with alternative approaches that are much slower in realizing their goals. As dairying is perceived as the way out of poverty, many farmers accumulate the capital to start a dairy unit by purchasing and growing male calves from dairy framers, or use AI to upgrade their local animals. Due to feed shortages, male calves, which are considered of low value in intensive dairy farms, are sold off immediately after weaning at a price of approximately KSh. 3,000 (US$ 43). After 2-3 years of feeding the male animals are sold to butchers for between KSh. 25,000 – 30,000 (US$ 357 – 429). Through this system poor households can purchase a dairy heifer to start dairy production. On the other hand, poor farmers who cannot accumulate enough start-up capital grow forage crops like Napier grass for sale to dairy farms. Cereal crop residues are also sold or exchanged for manure.

More than 80% of the population in Africa lives in the rural areas and is engaged in agricultural activities. The level of poverty is much higher in the rural, compared to the urban, population. As this is the population that keeps livestock, the livestock revolution should improve their incomes. However, small-scale, rural producers may benefit less than the emerging, larger-scale producers in peri-urban areas (a phenomenon already seen in S.E. Asia) unless there are significant investments to promote the supply side in terms of improving rural infrastructure and institutions. Bottlenecks in these areas represent enormous challenges in attaining food security and equity in favour of rural smallholders. It is becoming increasingly clear to many agriculturalists in Africa that much larger financial commitments will be necessary to achieve longer-term growth for African agriculture and its allied sectors, education and health. For example, subsidies or market stabilizers that were abandoned during implementation of structural adjustment programmes in the last decade, need to be reintroduced to enable smallholder agriculture to intensify and to enhance its competitiveness (Jayne *et al.*, 2002). The question is whether African governments can afford the high costs that must be incurred for such support. Without this kind of support, the potential benefits of the revolution may by-pass poor livestock keepers in Africa.

The livestock revolution and food security

Most small-scale farmers are mixed crop and livestock producers. The crop-livestock interactions in these enterprises create opportunities for synergy that improves farm productivity. Existing opportunities need to be further exploited. As most of the farmers use little, if any, chemical fertilizer, use of manure is very important in maintaining soil fertility.

Due to major soil erosion problems, the government of Tanzania ordered the withdrawal of all livestock from the eroded area of Kondoa Region in the semi-arid zone of central Tanzania in 1979 (Tekie *et al.*, 1996). Although livestock provided milk, meat and cash income to the farmers in the area, manure was the most important output and was used to sustain food crops yields from inherently infertile soils. After the re-introduction of livestock in the area in 1988 the soil characteristics improved and could support the production of food crops (Table 3.4)

Table 3.4. Chemical characteristics of soil in Kondoa Region after the introduction of goats.

Year	*Soil pH (1:2.5)*	*Total N (%)*	*Extractable P*	*Exchangeable bases c.mol/kg*			
				CEC	*Ca^+*	*Mg^+*	*K^+*
1996	5.75	0.12	4.34	10.7	4.26	1.7	0.76
1999	6.36	0.61	6.82	12.6	6.65	2.0	1.16
Difference	0.61	0.49	2.48	1.9	2.39	0.3	0.4

Adapted from Shirima and Msechu (2003)

Use of goat manure improved the yield of maize from 450 to 1,450 kg/ha and that of sorghum from 380 to 900 kg/ha due to the improved soil fertility (Shirima and Msechu, 2003). Therefore, households keeping goats in Kondoa Region had a better food supply when compared to families that did not keep goats. In Niger, tethering of animals on cropland for three days increased the pearl millet yields to about 180 kg in a plot 4 x 4 m compared to 10 kg when no manure was applied (ILRI, 1997). Mixed farming also allows conversion of low quality crop residues to valuable livestock products.

Considering that mixed crop-livestock systems predominate in Africa (McIntire, 1992), the intensification process will have to address both the crop and livestock components of the production, as a shift to a more specialized production system is not envisaged.

The livestock revolution and the environment

While many current small-scale peri- and intra-urban crop-livestock systems have beneficial interactions and synergies amongst farm enterprises (e.g. using crop residues as forages, and manure as fertilizer for crops), important challenges arise when specialist livestock enterprises predominate. In specialized livestock production systems in peri-urban areas, the enterprise depends on feed importation from the rural areas, as forage and from industrial areas as compounded feeds. This leads to an accumulation of nutrients in the peri-urban areas and disposal of waste becomes a major concern. The role and efficiencies of forage and manure markets, and the associated environmental impacts of changes in local and landscape nutrient availabilities, need to be evaluated, and policy options to reduce potential pollution and public health hazards considered. Regions with extensive production systems and the potential for increased animal product off-take to supply cities, and the conditions necessary for exploiting that potential, need to be identified. An increasingly important field of research will be the public health hazards of animal product marketing, market constraints and the development of appropriate local and regional standards and regulations for such markets. The focus will be on how impending changes in the consumption of animal products and their production can be exploited, and how the strategic lessons learnt can benefit the urban and rural poor, while ensuring the protection of the urban, peri-urban and rural environments.

Table 3.5. Trends in trade in livestock products.

Index of Net Exports	*Sub-Saharan Africa*		*Developing world*		*Developed world*	
	1993	*2020*	*1993*	*2020*	*1993*	*2020*
Beef (million MT)	0.0	-0.2	-0.4	-2.7	0.4	2.7
Milk (million MT)	-1.5	-1.8	-18.7	-30.9	18.7	30.9
Pork (million MT)	0.0	-0.1	0.7	0.2	-0.7	-0.2
Poultry (million MT)	-0.1	-0.2	-0.5	-1.8	0.5	1.8

Source: Delgado *et al*. (1999)

The livestock revolution and trade opportunities

Table 3.5 shows that net imports of livestock products to Africa will marginally increase. Currently, although some African countries export live animals to Middle East countries and Botswana exports beef to the north, the quantities relative to total production are relatively small. To increase these exports, the continent needs to address the issue of livestock diseases and provide production support and market stabilisation if it wants to increase exports of animal products. Since the resources to do this cannot be easily envisaged, markets will remain local.

Developed countries, especially the EU and North America, use subsidies and other trade barriers to protect their domestic markets. These subsidies have distorted the patterns of international trade and although the developing countries hold a large proportion of the world's livestock, they contribute only a relatively small proportion of the world trade in livestock and livestock products. Although the World Trade Organization (WTO) aims at reducing distortions caused by subsidies, the introduction of non-tariff barriers will continue influencing trade in livestock products. International standards when embedded into national standards will also affect cross-border regional trade. Therefore, the raising of standards can lead to public services and markets that discriminate against poor livestock keepers.

The other main problem is that the poor nations of Africa do not effectively participate in setting the standards; methods that improve their participation in setting appropriate trade standards should be developed (Holden, 2002). There is also a tendency to (unnecessarily) rely on international food safety standards without proper examination of the suitability of the standards for domestic markets. Reliance on strict international standards often relegates a large proportion of the domestic market to operating informally and without policy support. The trend suggests that either the countries do not have the capacity to set more appropriate local standards, or they fail to appreciate the fact that the international standards are based on bottom line risk thresholds, that are often higher than the national requirements of many countries, including developed countries, and that the Sanitary and Phyto-Sanitary (SPS) agreement under WTO explicitly permits governments to choose not to use the international standards. Indeed the SPS agreement itself encourages governments to establish national, bilateral or regional SPS measures consistent with, but not necessarily the same as, international standards, guidelines and recommendations. The SPS, therefore, allows for bilateral or regional agreements. An important issue for the livestock product markets of Africa therefore, is to adopt standards that do not unnecessarily hurt domestic and regional markets, especially where there are insignificant exports outside a region and, or, where sanitary conditions are similar. Undeniably, scientific evidence is required to avoid unnecessary trade restrictions or barriers supposedly based on sanitary concerns.

Conclusion

Africa has remained a net importer of livestock products (FAOSTAT, 2002) and with the projected increased demand the continent might remain a net importer for sometime. Importation of cheap livestock products might benefit the consumers who are mainly the urban rich but disadvantage the mainly rural poor livestock producers. On the other hand the livestock revolution can be exploited through adoption of policies that encourage trade and smallholder participation. This would have a major impact on poverty in the continent as the majority of livestock producers are in the poor strata of society. Though demand that can be exploited will remain local, the opportunities could have been improved through international trade. However, due to the subsidies and trade barriers imposed by the developed countries, Africa will not be able to increase her participation in international trade in livestock products in the foreseeable future. Unless there are major new investments in the development of infrastructure, relevant technology development and disease control, few benefits of the revolution will be realized domestically and regionally.

In addition, if the small-scale poor livestock producers are to benefit from the livestock revolution, then policies must be put in place that make sure that the benefits go to the poor producers and consumers as well and not only to multinationals, who, inevitably will set up industrial-scale, commercial production units similar to those appearing in parts of Asia.

References

Anderson, S., Rushton, J, Staal, S and Tulachan, P. (2002). Demand and supply side changes in the milk sector: impact on smallholders. In: *Responding to the increasing global demand for animal products - Programme and Summaries*. International conference organised by the British Society of Animal Science, American Society of Animal Science and Asociacion Mexicana de Produccion Animal, held at the Universidad Autonoma de Yucatan, Merida, Mexico, 12-15 November, 2002. British Society of Animal Production, Penicuik, UK, pp. xviii.

Anon (2003). *Kenya rural development strategy*. Government of Kenya. Nairobi, Kenya, pp. 72.

Bureau of Statistics (1996). *National sample census of agriculture 1994/5. Tanzania mainland report*. Volume II Ministry of Agriculture and Cooperative (MOAC). Bureau of Statistics, Planning commission, Dar-es-Salaam, March 1996.

Delgado, C., Rosegrant, M., Steinfeld, H., Ehui, S., and Courbois, C. (1999). *Livestock to 2020: The next food revolution*. Food, Agriculture and Environment Discussion Paper 28, International Food Policy Research Institute (IFPRI), Washington, DC, USA.

FAOSTATs, (1983-2002). *FAO statistics database on the World Wide Web*. http://apps.fao.org (accessed February 2003).

Gitau, G.K., O'Callaghan, C.J., McDermott, J.J.., Omore, A. Odima, P.A., Mulei, C.M. and Kilungo, J.K. (1994). Description of smallholder dairy farms in Kiambu district, Kenya. *Preventative Veterinary Medicine* 21:155-166.

Holden, S. (2002). How do trade regimes make livestock producers vulnerable? In: *Responding to the increasing global demand for animal products - Programme and Summaries*. International conference organised by the British Society of Animal Science, American Society of Animal Science and Asociacion Mexicana de Produccion Animal, held at the Universidad Autonoma de Yucatan, Merida, Mexico, 12-15 November, 2002. British Society of Animal Production, Penicuik, UK, pp. 4.

International Livestock Research Institute (ILRI) (1997). *The Backgrounder*, International Livestock Research Institute (ILRI), Nairobi, Kenya

Jayne, T.S., Minde, I.J and Argwings-Kodhek, G. (ed.) (2002). *Perspectives on agricultural transformation: A view from Africa.* Nova Science Publishers, Inc., New York, 277 pp.

Leeuw, P. de, Omore, A., Staal, S. and Thorpe, W. (1999). Dairy production systems in the tropics. In: *Smallholder dairying in the tropics.* Edited by Falvey, L. and Chantalakhana, C. International Livestock Research Institute (ILRI), Nairobi, Kenya, pp. 19-44.

McIntire, J., Bourzat, D. and Pingali, P. (1992). *Crop-livestock interactions in Sub-Saharan Africa.* World Bank, Washington, DC, USA.

Ndikumana, J., Stuth, J., Kamidi, R., Ossiya, S., Marambi, R. and Hamlet, P. (2000). *Coping mechanisms and their efficacy in disaster prone pastoral systems of the Greater Horn of Africa. Effects of the 1995-97 drought and the 1997-98 El Nino rains and the responses of pastoralists and livestock.* ILRI Project Report A-AARNET (ASARECA-Animal Agriculture Research Network), Nairobi, Kenya, GL-CRSP LEWS (Global Livestock Collaborative Research Support Program Livestock Early Warning System), College Station, Texas, USA and ILRI (international Livestock Research Institute, Nairobi, Kenya, pp.124.

Odima, P.A., McDermott, J.J. and Mutiga, E.R. (1994). Reproductive performance of dairy cows on smallholder dairy farms in Kiambu district, Kenya: Design, methodology and development of constraints. In: Proceedings of the 7th ISVEE symposium, 15 – 19 August, 1994. Nairobi, Kenya. *The Veterinarian* 18:366.

Omiti, J. and Staal, S. (ed.) (1996). The Ugandan dairy sub-sector: A rapid appraisal. ILRI/NARO/MAAIF Research Report, International Livestock Research Institute/,National Agricultural Research Organization (Uganda)/ The Ministry of Agriculture, Animal Industry and Fisheries (Uganda). International Livestock Research Institute (ILRI), Nairobi, Kenya, 137 pp.

Omore, A. and Staal, S. (ed.) (1998). *The Tanzania dairy sub-sector: A rapid Appraisal.* Volume 3, Main Report, International Livestock Research Institute (ILRI), Nairobi, Kenya.

Omore, A., Muriuki, H., Kenyanjui, M., Owango, M and Staal, S. (1999). *The Kenya dairy sub-sector: A rapid appraisal.* Smallholder Dairy (Research & Development) Project Report, 51pp.

Shirima, E.J.M. and Msechu, J.K.K. (2003). Zero grazing of goats for improved seasonal feed availability and soil fertility in Kondoa, Tanzania. In *Improved all-year-round feeding and marketing of livestock products.* Edited by Mwangi, D.M. and Kariuki, J. Proceedings of the Animal Production Society of Kenya (APSK)

Symposium held at the National Animal Husbandry Research Centre Niavaisha, Kenya 7-8 March, 2003. (in press).

Staal, S. J., Owango, M. Muriuki, H., Kenyanjui, M., Lukuyu, B., Njoroge L., Njubi, D. Baltenweck I., Musembi F., Bwana O., Muriuki, K. Gichungu, G., Omore A., and Thorpe W. (2002). *Dairy systems characterisation of the greater Nairobi milk shed.* Smallholder Dairy (R&D) project. Kenya Agricultural Research Institute (KARI)/Ministry of Agriculture (MoA)/International Livestock Research Institute (ILRI) Collaborative Dairy Research Programme, International Livestock Research Institute (ILRI), Nairobi, Kenya.

Tekei, G., Masaoa, A.P., Shayo, C.M., Olotu, H.A. and Shirima, E.J.M. (1996). *Zero grazing, an alternative system for livestock production in the rehabilitated areas of Kondoa Tanzania.* Regional Soil Conservation Unit (RSCU) and Swedish International Development Co-operation Agency (SIDA). Report Series No. 15, Nairobi, Kenya.

that all countries, both rich and poor, can benefit from greater specialisation in the products for which they have a comparative advantage. Free trade in international markets is therefore predicted to lead to a welfare optimising distribution of production and consumption between countries, with all commodities produced at minimum cost (where costs include those of transactions and transport).

Other advantages are claimed to result from trade liberalisation. Increased trade is expected not only to increase incomes and consumer choice but also to reduce unemployment and promote economic growth. International capital mobility may reduce the impact of differences in the capital resource base. A further presumed benefit of free trade is the spread of new technology from country to country. In countries shielded from foreign competition by trade barriers, traditional technology and methods of production are likely to hold sway. Under free trade, however, competition with foreign producers requires the domestic producers to stay abreast of the latest technologies. Furthermore, the general contacts made through trade contribute to the diffusion of modern technologies from the developed to the less developed countries.

Against these presumed benefits, critics of free trade argue that developing countries become less self sufficient and more dependent on other countries. The argument that countries dependent on primary sector exports are likely to suffer from deteriorating terms of trade, or falling relative export prices, is hardly relevant since developing countries are now net food importers and demand for livestock products is much more income elastic than demand for other foods. A further criticism is that trade becomes increasingly concentrated in the hands of trans-national corporations (TNCs). It should be recognised, however, that the TNCs are the main delivery route for both capital (foreign direct investment) and new technology to developing countries. These inputs can serve as catalysts for economic development. To benefit from them developing countries need to adapt technologies for domestic use, provide competitive policy environments and design contracts that allow small scale producers to thrive within the operations of TNCs, which is not easy to achieve. While intensive livestock keeping, pigs, broilers and some dairy enterprises may be integrated with processing and marketing plants, there are major problems in integrating traditional producers with large organisations. None the less the general view among economists is that all countries benefit, in the long run, from trade liberalisation (Bruinsma, 2003; World Bank, 2001).

Comparative advantage and product choice

Differences in comparative advantage between countries depend upon

(a) the available resources of land, labour and capital and (b) the available technology. The two are closely related both because capital resources are intimately involved in the development and implementation of new technology and because relative factor scarcity influences, and induces changes in, the prevailing technology.

The resource base is influential in that developing countries with plentiful land and labour resources, like much of Africa and Latin America, but with limited capital, have a comparative advantage in the production of land-using and labour intensive agricultural products. The fact that the Low Income and Least Developed Countries of the African Sahel export live ruminants, whilst importing other livestock and livestock products, probably reflects their comparative advantage in grazing livestock on the extensive semi-arid rangelands of that sub-region. Latin American countries such as Argentina and Uruguay are long-established beef exporters.

However, global demand for intensively produced poultry meat, and for pig-meat in some countries, is growing faster than the demand for ruminant meat. A contributory factor is the relatively low price of these meats resulting, in part, from the greater productivity of these species. Comparative productivity of the different livestock species, derived from FAO statistics, are presented in Figure 4.5. Production is measured by the number of animals/birds slaughtered in 2000 plus the increase, or minus the decrease, in animal/bird population between 2000 and 2001. Productivity is then estimated by number slaughtered as a percentage of the population recorded in 2000. This simple comparison serves to emphasise the greater reproduction rates of the 'landless' livestock species.

While these livestock enterprises require little land, intensive production and processing necessitate increased capital inputs and more sophisticated technology than production of grazing ruminants. This may shift the trade advantage to the developed countries, where these resources are more readily available and productivity is considerably higher. The higher productivity in developed countries is illustrated in Figure 4.5. The USA, which has the necessary capital and technology, now dominates the global export trade in poultry meat.

Similarly intensive milk production and processing generally yields higher returns than beef production, but requires more capital and higher levels of technology. The apparent comparative advantage of the developed countries in the production and export of dairy products is due to the greater availability of capital and the availability of modern dairy technology. It is noteworthy, however, that India is now the world's largest producer of milk.

for increasing domestic production and self-sufficiency, while the general expansion of trade may benefit other sectors of the developing country economies.

Trade integration and the European Union

Regional trading blocks can take the form of free trade areas, customs unions, common markets and economic unions in order of increasing integration between members. By reducing or eliminating trade barriers between members such organisations create trade, but only at the expense of diverting trade away from potentially cheaper sources in non-member countries. The European Union (EU) is one of the oldest and most strongly established trading blocks but another 150 such associations have been formed over the last 50 years and many still survive. Many of these other groups have had limited success in creating trade possibly because they share comparative advantage in the same products. However, 70% of EU trade is between member countries.

The Common Agricultural Policy of the EU has provided financial support and price stabilisation for European farmers, but this has been at the expense of raising barriers against imports from third countries and thereby depressing and destabilising world prices. In the past the European Community was guilty of "dumping" meaning the disposal of agricultural surpluses, resulting from producer subsidies, at artificially low prices. This practice was prevalent in the 1980s by which time the European Community had accumulated large stocks of beef and dairy products through intervention buying. Cheap sales of beef from the European Community to coastal West African Countries in the early 1990s covered a large proportion of their demand (60% of the beef supply for Ghana and 40% of that for Côte d' Ivoire) and caused a serious drop in trade from the Sahelian Countries to the Coast (Van Ufford and Bos, 1996). It should also be mentioned in this context that low priced exports of dried milk, from developed countries to India, provided the basis for the success of 'Operation Flood' in the development of the dairy industry. A decline in price support through intervention buying in the European Union, together with anti-dumping measures introduced under the General Agreement on Tariffs and Trade (GATT), has reduced the incidence of dumping. None the less the EU, along with the USA and other developed countries, maintains high tariff levels, especially for temperate-zone basic food commodities (see Figure 4.6).

Favoured treatment was offered to 70 African, Caribbean and Pacific Ocean (ACP) States under the Lomé Convention, recently replaced, as an interim measure by the Cotonou Agreement. The Lomé Convention provided for the stabilisation of export earnings and a

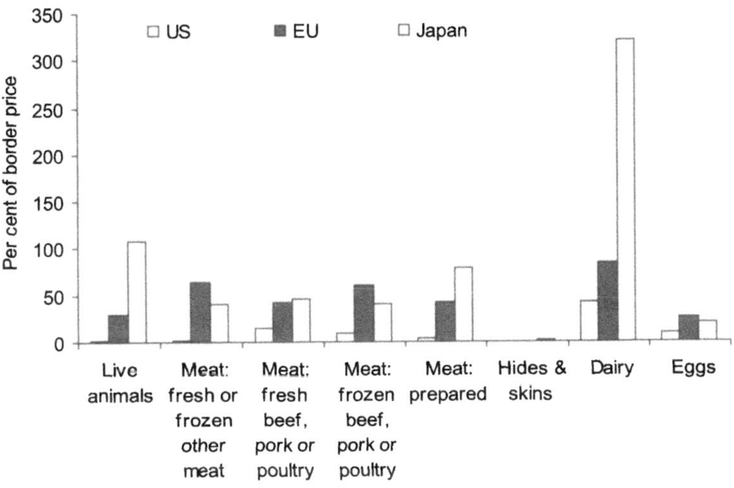

Figure 4.6. Average tariffs on livestock products in the US, EU and Japan.

significant reduction of tariffs on bananas, sugar, rum and beef and veal from the ACP countries. Botswana, Zimbabwe, Madagascar, Swaziland and Kenya have benefited from the beef protocol. The trade preferences to the ACP countries in particular must be abandoned by 2008 at the latest but the Least Developed Countries will benefit from free, un-taxed access to European markets for all products from 2005.

The WTO and developing countries

The General Agreement on Tariffs and Trade (GATT) launched in 1947 was aimed at trade liberalisation. It was not until the Uruguay Round of negotiations, from 1986 to 1994, that trade in the main temperate agricultural products was discussed. The Agreement on Agriculture, signed in 1994, called for reduction of export subsidies, reduction in financial support for agricultural producers and improved access through replacement of non-tariff barriers by tariffs which would be gradually reduced. The GATT was replaced by the World Trade Organisation (WTO), which had stronger powers, at the end of the Uruguay Round.

Major global economic benefits were predicted from the increased trade resulting from the Agreement on Agriculture and the establishment of the WTO. Most studies predict a resultant modest increase in food prices, which will benefit exporters but may exacerbate food security problems of Low Income Countries. Some authorities predict falling food prices however. Reduction of agricultural price support in Europe and the USA should lead to shifts in production and exports. Dairy exports may increase from Oceania, Eastern Europe and Southern Africa and from developing countries of South Asia and Latin America. Beef and lamb exports may expand from Oceania and Latin America. However it is predicted that Europe and North America may increase pig and poultry meat exports. Brazil, Thailand and Hungary are also predicted to increase exports of poultry meat. Developing countries, as a group, face problems not only of rising prices of food imports, but also of meeting the costs of compliance with the recommended customs and trade procedures recommended by the WTO (Finger and Schuler, 1999).

Hitherto, there has been limited progress in reducing protection of domestic producers. The Doha Round of WTO negotiations, launched in November 2001 has served to demonstrate the wide diversity of levels of commitment to trade liberalisation. However, there is general agreement that 'Special and Differential Treatment' should be applied to developing countries, requiring less demanding policy changes on their part. It does little to reduce the impact of producer support in the developed countries.

Although policies in the European Union and North America have shifted away from trade distorting price supports, the overall level of support remains high, as shown in Figure 4.6. Indeed critics have suggested that the AOA has 'institutionalised' the production- and trade-distorting policies of the developed countries, without addressing the fundamental concerns of the developing countries (Green and Priyadarshi, 2002). Other commentators emphasise the damage done to developing country producers by the agricultural and trade policies of the developed countries (Binswanger and Lutz, 2000). A contrary view is expressed in Bruinsma (2003), Chapter 9, where a review of studies and assessments of the likely impacts of the dismantling of agricultural subsidies by OECD countries, leads to the conclusion that the impacts would be limited. It is argued that farmers in developing countries would benefit more from domestic policy reforms.

Regulatory measures

The SPS Agreement

Trade in agricultural produce carries the risk of transmitting plant or animal pathogens and human food pollutants across national boundaries. Importers need assurance that adequate quality and safety standards have been met to protect the life and health of their people, plants and animals. Assurance can be provided by either a system of border inspections and controls, with the possible rejection of consignments not meeting the required standards, or evidence that production, in the country of origin, is in accordance with agreed quality and safety standards. In either case, some of the costs of risk limitation, by the importer, are passed on to potential exporters. Sanitary and phytosanitary (SPS) issues are a common cause of trade disputes, seen by some developing countries as disguised measures of protection. Countries of origin are therefore entitled to expect safety and quality standards to be transparent, science-based and no more trade restrictive than necessary to meet the stated objective.

The SPS Agreement of the WTO (1995) is aimed at harmonising the health and safety standards applied internationally in line with the recommendations of the International Office of Epizootics (OIE), the International Plant Protection Convention (IPPC) and the *Codex Alimentarius*. This involves recognising the equivalence of different measures giving the same level of protection, allowance for adaptation to regional conditions, use of risk assessment to establish the appropriate level of protection and establishment of a formal framework for consultation and dispute settlement.

While consumers and producers in developed countries demand high food quality and safety as well as animal and plant health standards, standards in the developing countries are usually lower. Implementation of animal and plant health and food safety standards is constrained by resource limitations in general, including deficiencies in infrastructure, technology and skills. For nations not seeking to export to the developed countries, the existing level SPS standards may be all that can be justified economically. However, a strong case can generally be made for policies aimed at improving hygiene and the health of livestock, crops and humans. Obvious economic benefits are derived from improvements in animal and crop health, while food safety contributes directly to human welfare. There are, of course, external benefits to the developed countries, of such measures, in terms of the reduced risks of introduction of disease from developing countries.

Potential exporters among the developing countries face substantial costs in meeting the rising SPS standards required by the developed countries. Of the respondents to a survey of African food exporters in the mid 1990s, 57% stated that exports had been rejected, following border inspection, within the previous two years. Internal facilities for testing and inspection before export were inadequate due to financial constraints (Mutasa and Nyamandi, 1998). Participation in international committees, on technical and legal aspects of SPS control, is limited, while efforts at improvement are often hampered by the small-scale of the export trade. The costs of meeting SPS requirements for exporting livestock and other products to developed countries are very high. For Argentina to meet SPS requirements on meat, fruit and vegetables cost US$82.7 million between 1991 and 1996 (Finger and Schuler, 1999). Upgrading slaughterhouses in

Appendix 4.1. Developed and developing countries

The broad classification of countries into developed and developing categories, is widely used in UN publications. (A full listing is given in Bruinsma, 2003). The developed countries comprise (a) the high income, industrial countries of Western Europe, North America, Oceania, Israel, Japan and South Africa and (b) the transition economies of Eastern Europe and Central Asia. The former group of industrial countries broadly coincides with membership of the Organisation for Economic Co-operation and Development (OECD).

The developing countries include the mostly low income countries of Sub-Saharan Africa, low- and medium income countries of Latin America and the Caribbean, low-, medium- and high-income countries of the Near East and North Africa and low- and medium-income countries of South Asia and East Asia. Thus, there is a wide range of average *per-capita* income levels within the developing country category. Two important sub-classes of the developing country category, namely the 'least developed countries' and the 'net food-importing developing countries' are subject to special trade concessions under the WTO.

5

Disease and biosecurity constraints to trade in animal products

A.D. James

Veterinary Epedimeology and Economics Research Unit, School of Agriculture Policy and Development, The University of Reading, Earley Gate, PO Box 237, Reading RG6 6AR, UK

Opportunities for international trade in animal products

The major centres of consumption of animal products are countries of relatively high *per capita* income and human population density. In these countries, the cost of production of animal product tends to be high. In particular, land and labour prices are higher than in less populous and wealthy countries. This also tends to encourage the adoption of more intensive animal production systems, often associated with adverse effects on animal disease risk, product quality, animal welfare and the environment. These effects all increase the true economic cost of animal production, even though much of it is not borne by producers. Consumers in wealthier countries are apparently becoming increasingly concerned about these external costs, as evidenced by rapid increase in the demand for organic products and the widespread adoption of farm assurance schemes by food retailers.

There are very large differentials between different countries and regions in the prices of animal products. The existence of these price differentials is due to various constraints on trade in animal products. Without such constraints to trade, price differentials would be expected to reduce to a point where they reflected only transport costs and consumer preferences. Reducing constraints on international trade would be to the overall economic benefit of both exporting and importing countries. Although there would be some shifting of economic welfare from consumers to producers in exporting countries, and from producers to consumers in importing countries, it would in principle be possible for government policy to produce an equitable redistribution of the national benefits.

Constraints to trade in animal products

Constraints to trade in animal products are many and complex. Some are due to real problems such as perishability and disease risks and others to agricultural support policy in potentially importing countries. In principle, agricultural support policy would be implemented by tariff barriers, while other constraints would be considered as non-tariff barriers to trade. However, non-tariff barriers are often motivated, at least in part, by agricultural support policy.

Many livestock products are highly perishable and cannot be transported over long distances at reasonable cost without deteriorating. This is especially true of bulky products such as fresh milk or animal carcasses. However, these products can be processed to extend their life and/or reduce their bulk. Milk can be heat treated or dried, and meat may be de-boned and frozen. Such processing may reduce the market value of the products. However, processing techniques are improving and transport costs becoming lower. Therefore constraints of bulk and perishability are much less important than they once were, and likely to become even less so in the future.

Agricultural support policy in many countries includes maintaining product prices at artificially high levels by restricting imports. The restrictions may be in the form of tariffs on imports, but may also include non-tariff barriers based on food safety and animal health issues. Whether or not these are intended to be part of agricultural support policy, their effect is to constrain imports and increase product prices.

Failures of producers in exporting countries to meet the animal welfare and environmental protection standards demanded of producers in importing countries have been discussed as possible reasons for restricting imports. As public concern over such issues increases, it is more likely that they will be used as a reason for restricting imports.

Most countries importing animal products for human consumption demand strict food safety standards. It can be very expensive to meet these requirements, and they would certainly prevent many potential exporters from gaining access to some markets.

Animal disease control restrictions are probably the most widely-applied and onerous constraint to international trade in animal products. These affect products that are not for human consumption, such as hides and skins, as well as food products.

Food safety risks

Food safety risks in international trade in animal products include zoonotic diseases (diseases that can be passed from animals to man), other food-borne pathogens and residues of drugs or other chemicals that may be contained in animal products. These risks can all be managed at the level of slaughter and processing plants. Most importing countries will only accept animal products for human consumption from plants that they have inspected and approved. The inspections cover the operating procedures, as well as buildings and equipment. Some smaller importing countries will accept products from plants approved by the European Union (EU) or the United States Department of Agriculture (USDA) without conducting their own supervision. The major operational requirements are usually the operation of meat inspection to detect zoonotic diseases, a quality assurance scheme, Hazard Analysis Critical Control Points and monitoring for food-borne pathogens and residues. While these requirements are demanding and expensive, they can be met by any modern and well-managed slaughter or processing plant. Due to the high capital investment and human resource requirements, only large scale plants would be able to achieve the required food safety standards at reasonable cost, but small-scale producers would find it difficult to participate in international trade for marketing and other commercial reasons. Food safety requirements are necessary to protect public health, not unduly expensive and cannot be seen as an unwarranted constraint to trade.

In response to the crisis in consumer confidence due to bovine spongiform encephalopathy, the EU demands complete traceability of bovine livestock from its own producers. Many beef-exporting countries would be unable to achieve this due to the nature of their production systems, and the EU has not yet required such comprehensive traceability for beef imported from third countries. If such requirements were to be imposed in the future, most exporting countries would be able to achieve traceability in part of their industry, in much the same way as they maintain 'export' and 'non-export' abattoirs.

Animal disease risks

The principal animal disease concern in international trade is the introduction of epidemic diseases such as foot-and-mouth disease. Some of these diseases are readily introduced in animal products, and have the capacity to spread very rapidly. The economic cost of dealing with outbreaks can be very high. The costs of the 2001 foot-and-mouth disease (FMD) outbreak in the United Kingdom were officially estimated to be £ 8 billion.

Animal disease agents can be introduced by legal or illegal trade. The risks in legal trade can be managed by various controls, and these have been very successful in preventing the introduction of animal diseases to importing countries. Illegal trade, ranging from individual travellers carrying small quantities of animal product to semi-commercial importation of speciality products, is not subject to any animal health or food safety controls. The only means of managing the risks arising from illegal trade is through intensive controls at entry points. These are very difficult to impose without causing undue disruption to *bona fide* trade and travellers. Illegal trade remains, therefore, the most important and uncontrolled risk of introducing exotic animal diseases.

Various measures have been used to control animal disease risks in legal trade. These have been very successful in preventing the introduction of major epidemic diseases to importing countries.

The most basic measure to control animal disease risks is the control or eradication of the OIE List 'A' diseases (those deemed to be of greatest significance in international trade) in exporting countries. For example, most importing countries would only accept meat from countries either free of FMD or having zones free of the disease. Although this can be extremely expensive for exporting countries, the animal production benefits of controlling FMD almost invariably outweigh the costs. However, in some situations it is not feasible to eradicate certain diseases. There may be a reservoir of infection in wildlife, or infection in neighbouring countries. In these situations, exporting countries have to maintain a vaccinated zone, an unvaccinated buffer zone with intensive surveillance and an export zone. Movement of animals between zones must be strictly controlled. These measures are very expensive, both in direct costs and through their impact on wildlife, the environment and normal patterns of animal production. Maintaining populations of unvaccinated animals close to reservoirs of infection carries risks of epidemics spreading through unprotected animals in the buffer and export zones. For many potential exporting countries the costs and risks are prohibitive, and these countries are excluded from international trade.

Inspections in export slaughter plants are a second line of defence to protect importing countries. Animals should be inspected before entering these facilities to ensure that they come from an area free of List 'A' diseases and show no clinical signs suggestive of infection. *Post-mortem* inspections must also confirm the absence of suspicious lesions. These inspections are also essential to control the risks associated with zoonotic diseases. Their effectiveness in controlling the risk of epidemic animal diseases is limited, in that animals could pass through the slaughter plant after infection, but before exhibiting any clinical signs or lesions. However, in this case, the disease would be expected to manifest itself in the animal population long before any products could reach importing countries. For this reason, most importing countries insist on identification and traceability of animals and products, at least from the farm despatching animals to the export slaughter plant, to the export consignments.

A third set of measures, processing techniques may also reduce or eliminate the possibility of viable disease agents surviving in animal products. For example, FMD virus does not survive the acidification of muscle tissue that occurs during the development of *rigor mortis*. Therefore, some importing countries will accept beef from countries not free of FMD, provided that it is free of bone and lymphatic tissue and that the carcass has been held for 24 hours to allow the development of *rigor mortis* before cutting and freezing. Various forms of heat treatment are also very effective in eliminating viable disease agents from various animal products, including milk.

Conclusions

Illegal trade in animal products presents much greater risk of introducing exotic animal diseases than legal trade for most importing countries. Some importing countries, such as Australia, impose far more stringent border controls than others such as EU countries. More research is needed to evaluate whether the cost of intensified border controls would be justified by reduced risk of exotic diseases. It will never be possible to eliminate exotic disease risks arising from illegal importation of animal products, and countries at risk need to examine the cost-effectiveness of border controls, compared to disease surveillance for early detection of exotic diseases and biosecurity measures (e.g. movement restrictions) to limit their spread.

Of the three groups of measures required by importing countries to control the risk of introducing exotic animal diseases, the disease status requirements are by far the most onerous and expensive for exporting countries. Slaughter plant inspections and product processing carry very little additional cost.

Many potential exporting countries are effectively excluded from international markets for animal products by the disease status requirements. Others bear extremely high financial, social and environmental costs in maintaining export zones. This restriction of trade is economically detrimental to both exporting and importing countries. In many cases, slaughter plant inspection and product processing procedures are much more cost-effective in reducing animal disease risks associated with international trade, especially when coupled with effective disease surveillance and product traceability.

It is considered that in many cases importing countries are over-cautious in the disease status requirements that they impose on

commercial farmers, veterinarians, the Department and Sub-Departments of Animal Health, animal health workers, exporters, veterinary drug and feed shop owners, and province and district level officials. The second phase took a year. Six suspected CSF outbreaks were investigated using participatory methods. Further key informant interviews were used to prepare enterprise budgets for farmers and traders. A pig population and market model was developed in Excel™ to assist with calculations of benefits and costs. In the third phase, the preliminary results were presented to the stakeholders originally consulted and the model inputs revised. Throughout the study, secondary data were obtained from a review of official statistics, government working papers, SVSV technical reports, reports of surveys by other projects, and grey literature.

Figure 6.1. Red River Delta.

Source: http://www.vacationsinvietnam.com/index-main.html, accessed 10/07/2003

Results

Importance of pigs to smallholder farmers

The national pig population is estimated at around 22 million. 70-80% of farmers keep pigs and small farms make up about 70% of the national farm population (IFPRI/DANIDA/MARD, 2000). Pig production can contribute as much as 50% to household cash income (Anh, 1999). Most large commercial farms have a complete system from piglet breeding to the finished pig, but smaller farms usually specialise in either sows or fattening pigs. Lan (2000) found that smallholder pig production can produce gross margins of VND 467,000 per sow per year for weaner production, and VND 16,000 or 23,000 per pig for grower and finished pig production respectively (US$1 = approx. VND15,000).

The pig market

Piglet marketing channels are shown in Figures 6.2 and 6.3. The proportion by volume and value through each channel varies with seasonal demand. There are different kinds of pig traders in different provinces, depending on the local livestock production systems. For example, in Thai Binh and Nam Dinh, the number of piglets produced in the breeding herds exceeds internal demand and they are sold to the south of Vietnam or exported. Only a few licensed slaughterhouses are permitted to export. In Ha Tay, very few piglets are sold out of the province, and there are many finishing farms producing for the Hanoi market. Hence, piglet wholesalers do not purchase piglets in Ha Tay. In all provinces, finished pigs may be sold directly to slaughter points or through finished pig traders.

Figure 6.2. Piglet marketing chain for provinces that export.

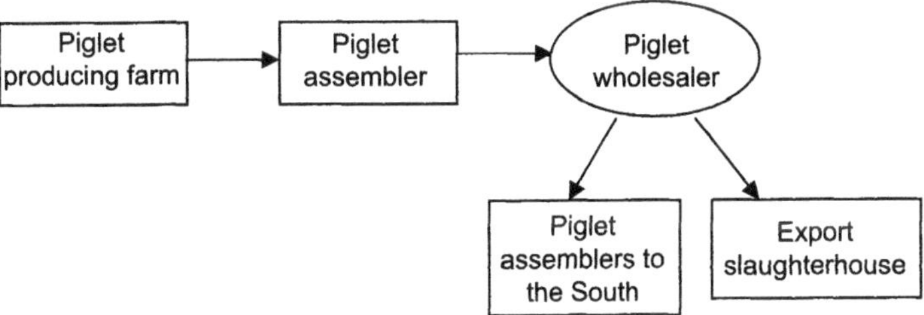

Notes:
Assembler = a small scale trader who collects piglets from smallholders and carries them by bicycle or motorcycle to a wholesaler.
Wholesaler = a large scale trader who transports 100-500 pigs at a time by truck and may buy from an assembler or directly from a commercial farmer.

Figure 6.3. Piglet marketing chain for provinces with a mainly local market.

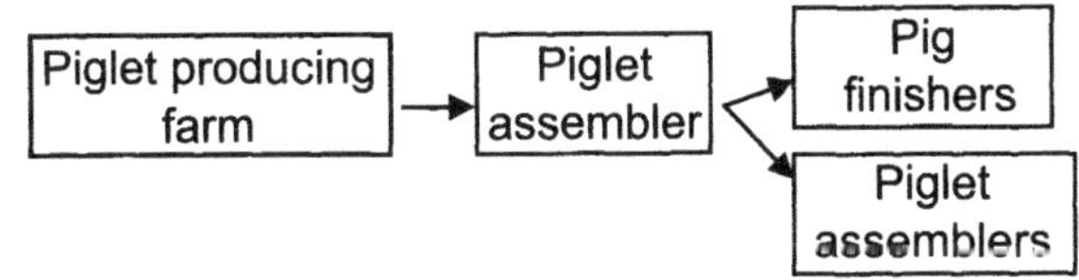

Notes:
Assembler = a small scale trader who collects piglets from smallholders and carries them by bicycle or motorcycle to a local market.

There is a huge demand for live piglets in the South of Vietnam and for export to Cambodia. There is also a demand for piglets to export as frozen carcases to other places, particularly for Chinese New Year celebrations. In the years 2002 to 2003, domestic demand was very high and domestic prices were higher than export prices, so some piglet slaughterhouses were forced to close temporarily. In other years, export prices have exceeded domestic prices and export trade has been more profitable. Smallholders and small traders make an important contribution to the niche suckling pig market. The export market for finished pigs is dominated by commercial producers.

Principal export markets are Hong Kong and Russia, with smaller amounts to Malaysia and "unofficially" to China. Competition from China (in Hong Kong), the EU, USA and Brazil (in Russia) is fierce. Current markets demand that slaughterhouses are certified and inspected for hygiene and meat tested for hygiene and quality (antibiotic residues and food-poisoning bacteria). Pigs must come from areas with no current outbreak of a List A disease (OIE list of highly infectious livestock diseases for which international notification of outbreaks is required), but none of the present markets requires national disease freedom. Foot and Mouth Disease occurs only occasionally and outbreak control is taken very seriously - an outbreak closes exports from a province. CSF is acknowledged to be widespread at low levels in a "chronic" form, and an outbreak requires quarantine in the surrounding area. Commercial farmers tend to protect themselves by fencing, purchasing animals from known sources and ensuring fairly good vaccination coverage, but smallholders do not.

Options for export expansion and potential impacts on smallholders

The national plan sets an annual export target of 40-50,000 tons of pig meat, but this has never been achieved (Table 6.1). The current export market is small, peaking in 2001 at approximately 27,000 tonnes of frozen and chilled pig meat.

Table 6.1. Estimated pork exports from Vietnam.

	000 tonnes	
	Ministry of Agriculture and Rural Development statistics	*Vietnam Economic News*
1988	1.7	
1989	5.1	
1990	9.5	
1991	15.0	25.0
1992	7.1	12.1
1993		19.7
1994		12.6
1995		6.4
1996		6.4
1997		1.4
1998		
1999		"inconsiderable"
2001	27.0	26.4

The following options could be considered for increasing export:

1. Reduced production prices to compete better in existing markets. Chinese exporters compete heavily on price, while Vietnamese producers face high feed costs and few subsidies.
2. Diversification of products e.g. into processed meat, adding value per kg shipped.
3. Diversification of markets for present products. Logical targets might include other ASEAN countries or Japan.

Option 3 might potentially be the most disruptive to smallholders and small traders. Entry to the WTO would be the first step towards market diversification, but would need to be followed by individual agreements with importing countries. To diversify markets, a two step process is likely to be required: initially, entry to the WTO, and then separate agreements with each new trading partner. Each step has implications for farmers and small traders, and these will now be explored in turn.

WTO entry

Entry to the WTO may be conditional upon a demonstrable ability to detect and control outbreaks of List A diseases. Vietnam's established disease recording system is mostly passive. An updated information system for outbreak detection and recording is being tested in a few provinces but requires further development (Mcleod *et al.*, 2003).

An increased presence of veterinary staff at commune levels might be considered a bonus to farmers. There are few fully qualified veterinarians in the private sector and it is accepted that government staff will engage in private practice in their free time. However, outbreak detection may result in the need to impose stricter movement controls, which would be considered an interference with domestic trade by producers and local traders.

Market diversification

Individual importers can be expected to impose SPS (Sanitary and Phytosanitary) regulations, under the general guidelines of the OIE and within the spirit of WTO intentions. They may in addition impose TBT (Technical Barriers to Trade) measures. Since the establishment of the WTO, in the place of the tariffs formerly imposed by some importing countries, there have developed an increasing number of technical regulations under the TBT and SPS agreements signed by member countries of the WTO at the end of the Uruguay Round in 1994. TBT measures are defined by Roberts and De Kremer (1997) as "Standards governing the sale of products into national markets, which have as their *prima facie* objective the correction of market inefficiencies stemming from externalities associated with the production, distribution and consumption of these products." Products failing to meet safety, quality or environmental standards can thus be excluded from import. SPS measures are a subset of TBT measures that are concerned with human and animal health risks from toxins, additives, contaminants in food and from plant and animal diseases. Roberts (1997) explains that "the TBT Agreement covers all technical regulations and conformity assessment procedures, except when these are sanitary and phytosanitary measures as defined by the SPS Agreement". A measure which proscribes use of an additive might be adopted to safeguard human health (an SPS measure) or to ensure the compositional integrity of a product (a TBT measure). TBT measures are based on national and international standards and include technical product standards, conformity assessments and other legal requirements (e.g. labelling regulations, packaging material types, weight measures and permitted ingredients). Unlike the TBT Agreement, the SPS Agreement requires that SPS measures are based on an assessment of risks posed by the import and provide an "appropriate" level of protection i.e. one which does not unjustifiably vary from levels of health or environmental protection provided by other measures.

Measures expected of Vietnam related to SPS regulations may include any of the following:

- certification of its veterinary service (for each new importing country);
- certification / inspection of export slaughterhouses (this is already done, but different standards may be required);
- traceability of livestock entering the export chain (currently not strictly enforced);
- inspection or certification of production premises (currently not required);
- strict quarantine of disease outbreaks and/or stricter routine movement controls.

It is possible that these measures could be imposed within a limited "disease free zone" or across the whole country.

For small farmers and traders within the disease free area, consequences might include:

- stricter trader licensing, raising entry costs for small traders, eventually excluding individual small traders from export market;
- costs of inspection creating a barrier that prevents very small producers from participating in the export process – this could particularly impact upon piglet producers;

- temporary disruption of trade through imposition of movement controls, and lost income for producers, traders and feed suppliers – this might result in development of a "grey" market, or vulnerable smallholders might go out of business;
- temporary disruption of production systems where piglets are normally transported from one area to another for fattening;
- in the longer term, lower disease levels should create increased income for all producers.

TBT regulations are likely to be concerned with:

- leaner carcasses (the lean content is currently 28%, the requirement in some markets is for 40%);
- heavier carcases (currently 29 kg per frozen half carcase, the requirement is usually for 40+ kg);
- facilities required in export slaughterhouses that can process at least 200 t/month (currently many process less than 150 t/month).

Possible results of these TBT regulations are:

- the need to produce exotic and crossbred pigs for the export market – these are less fertile and more prone to disease than local breeds, harder to obtain from local sources, and have higher input costs;
- closure of small export slaughterhouses reducing access;
- small traders banding together to supply larger batches on a consistent basis.

Conclusions

Over time, the regulations required to expand export markets are likely to change the shape of pig production and marketing. Eventually, small traders and smallholders in their present form may be excluded from the export market because of the technical requirements (need for exotic pigs) and sanitary requirements (process inspection and certification).

Smallholders will then have three options:

- to change to other livestock – poultry are a possible alternative, but pigs are very popular and considered profitable – the dairy sector is very small although its development is being explored by the government with external technical assistance;
- to focus on the domestic market, which is currently estimated to be expanding at over 8% a year – this has considerable potential, although the movement limitations imposed by disease free zones could create disruption;

Thirteen *campesino* households, belonging to the sector of "*La Era*" in SPT, participated in the study, with a mean arable surface of 2.70 ± 1.99 ha/family, with a range from 0.5 ha to 7.00 ha per household and a mean of 4.8 members per household. Monitored farms in San Pablo Tlalchichilpa were selected such that a range of farms in terms of size and farming practices were represented. In La Concepción Mayorazgo (LCM), twelve households participated in this study, with a mean farm size of 1.47 ± 1.10 ha and also a mean of 4.8 members per family. The monitored farms in LCM were those that have been participating in other studies (Chávez-Mejía *et al.*, 1998).

Information was gathered on changes in animal inventories and all outputs and income from livestock, whether in cash or in kind. Farmers were asked to ascribe an opportunity value to the produce consumed by the family, in terms of what they would pay to purchase the same product (Ayalew *et al.*, 2001; Wiggins *et al.*, 2001).

Ejido San Cristóbal

The second study area, the village of Ejido San Cristóba*l*, lies 35 km North-West of the city of Toluca, located at 19° 24′ N y 99° 51′ W, with an altitude of 2650 m, and a sub-humid temperate climate with rains in summer, with a mean temperature of 15.0 °C and 800 mm of total precipitation. Ejido San Cristóbal has 72 households of which 54 farm 189.75 ha of rolling slopes with limited irrigation (farm size between 0.5 and 11.0 ha). Forty seven households have small dairy herds (1 - 22 cows plus replacements) of upgraded Holstein-Friesian cows which are milked by hand twice a day, and milk is sold to local buyers who re-sell it raw in the city of Toluca (Arriaga-Jordan *et al.*, 1999a).

Feeding is the main component of variable costs, and the highest cash expenditures in these systems are on purchasing concentrates. Therefore, improved, lower cost feeding strategies were identified as a priority by *campesinos* at the project workshops (Arriaga-Jordan et *al.*, 1999a).

Traditional dry season feeding practices are based on maize straw and concentrates and in the rainy season rough grazing of uncultivated land, cut and carry forage and concentrates. Given the temperate conditions, and access to some irrigation, temperate pastures have been tried in cut and carry systems. The evaluation of grazing of cultivated pastures of ryegrass (*Lolium perenne* and *L. multiflorum*) associated with white-clover (*Trifolium repens*), integrated with maize silage in the dry season, comprises one of the main objectives of the project. This has been evaluated on-farm with cooperating farmers who have changed their traditional feeding strategies, by incorporating, under different intensity, the introduced pastures, grazing, and maize silage (Arriaga-Jordán *et al.*, 2002a).

After favourable results, and given the interest of, and in consultation with the community, a demonstration module was established in the school plot in 1999, consisting of 1.5 ha of perennial ryegrass / white clover pasture which is continuously grazed by a herd of six local upgraded Holstein cows, incorporating maize silage in the winter dry season (from 1.0 ha of bought maize fields) and between 3.0 and 4.0 kg of concentrates per cow per day, half the amount of concentrates usually fed by farmers. A cooperating farmer manages the demonstration module in close collaboration with the research team, so that it is representative of local management, which has enabled further experimentation and close monitoring of economic variables (Arriaga-Jordán *et al.*, 2002b)

The economic analysis of these evaluations has been undertaken by preparing farm budgets, from where summary measures are derived (Dillon and Hardaker, 1993). This method has been used successfully in the economic analysis of *campesino* dairy production systems in Mexico (Wiggins *et al.*, 2001).

Results and discussion

San Pablo Tlalchichilpa and La Concepción Mayorazgo in San Felipe del Progreso

There were twelve species of domestic animals identified in the study: horses (*Equus caballus*), donkeys (*Equus asinus*), mules (*Equus asinus X Equus caballus*), cattle (mostly *Bos taurus* but also some *Bos indicus X Bos taurus* crosses), sheep (*Ovies aries*), turkeys (*Meleagris gallipavo*), chickens (*Gallus gallus*), domestic European rabbits (*Oryctolagus cuniculus*), ducks both of the Peking type (*Anas platyrhynchos*) as well as muscovy ducks (*Carina moschata*) and their crosses, pigeons (*Columba columbae*), and pigs (*Suis scrofa*). Of these, the main species kept by the surveyed households were cattle, equines, sheep, and poultry (turkeys and chickens). Some households also keep other species, such as rabbits, ducks, pigeons, or pigs. The latter are not common in the hill slope *campesino* systems of San Felipe del Progreso. One household also breeds canaries, which are sold. Figure 7.1 shows the interactions of livestock in these *campesino* households.

Animals provide a number of economic benefits in cash or kind to the families (Table 7.1 and 7.2). All households in SPT obtained a benefit from animals that could be quantified. Four of the twelve participating households in LCM did not perceive any returns in cash or kind from their animals in the monitored period.

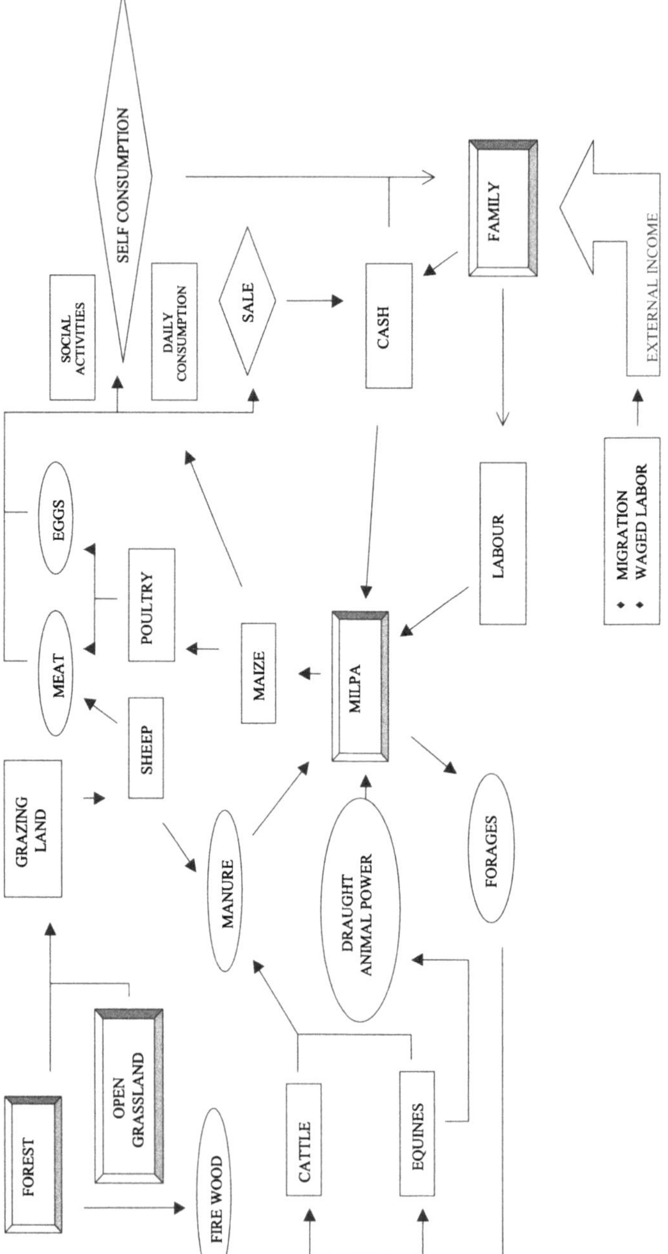

Figure 7.1. Interaction of livestock within the *Mazahua campesino* household.

On a per farm basis, 77% of income was received in cash in San Pablo Tlalchichilpa and 88% in Mayorazgo.

Total gross income for each village in cash or in kind is presented in Table 7.1. The proportional contribution per animal species in SPT to

Table 7.1. Total economic contribution of animals in San Pablo Tlalchichilpa (SPT) and La Concepción Mayorazgo (LCM).

	Cash		In kind		Total	
	Mex.	US $	Mex.	US $	Mex.	US $
San Pablo Tlalchichilpa	59,047	6,349	17,868	1,921	76,914	8,270
	77%		23%			
La Concepción Mayorazgo	22,140	2,381	2,910	313	25,050	2,694
	88%		12%			

Mexican peso (Mex.) 9.30 / US Dollar ($)

Table 7.2. Economic contribution of animals to *campesino* livelihoods in San Pablo Tlalchichilpa (SPT) and La Concepcion Mayorazgo (LCM) (July 1999 - June 2000).

	Total (Mex.)	US $	Equids	Cattle	Sheep	Poultry	Rabbits	Cash	In kind
SPT Total	76,914	8,270							
SPT Average	5,916	636	24%	56%	9%	10%	1%	77%	23%
LCM Total	25,050	2,694							
LCM Average	2,088	224	12%	60%	21%	7%	0%	88%	12%
Total	101,964	10,964	21%	57%	12%	9%	1%	80%	20%

Mexican peso (Mex.) 9.30 / US Dollar ($)

income from animals is: Cattle 56%; Equids 24%; Sheep 9%; Poultry 10%; and Rabbits 1%. In LCM the proportional contribution by species was: Cattle 60%; Equids 12%; Sheep 21%; and Poultry 7%.

Overall, farmers perceiving economic returns from their animals reported these as 80% in cash and 20% in kind. This is so because *campesinos* had difficulty ascribing opportunity costs to the benefits in kind they obtain from their animals. For example, in farm SPT12 which is the largest farm monitored (7.0 ha) and where the family relies heavily on their mule team for ploughing their land, the farmer basically did not ascribe any opportunity cost to the use of their mules. He insisted that there was *"no cost"* in using his mules; it was precisely because he owned his pair of mules that he was able to save on any "cost" since he did not have to pay rent.

On average, participating farms obtained Mex. 4,078 (US$ 438) from their animals in the monitored year. When the four households that did not record any economic benefit from their animals are not considered, the average benefit increases to Mex. 4,855/farm/year (US$ 522).

The gross income obtained on average is equivalent to 0.38 of the running minimum wage in the area for the period of study, with a wide range of economic contributions from the sale or consumption of a few chickens, to the sale of a bullock (LCM23), or the consistent income from the sale of milk (SPT4, SPT5, SPT6, and SPT11).

The mean annual income obtained from cattle at Mex. 6,149 (± Mex. 4,792) in the seven households from SPT who reported income from this species, considering that the income from LCM23 was extraordinary, was equivalent to 0.48 of the running minimum wage. For the four households in SPT who sold milk, the mean daily income from cattle (considering sales of milk plus livestock) was Mex. 21.63/ day, equivalent to 0.63 of the minimum wage which unskilled labour obtained in non-agricultural jobs in the area.

Work in other more favoured parts of the highlands of Central Mexico, has shown the viability of small scale dairy production to improve incomes of *campesino* families and make a better use of their limited land resources (Wiggins *et al.*, 2001). The conditions in San Felipe del Progreso, with a prolonged dry season and no irrigation available might not be conducive to efficient small-scale dairy farming, but the results obtained by the four households in SPT who sell milk warrant a more detailed study of their system as a possible option.

On the other hand, the only source of economic return from animals in eight of the 21 households perceiving economic returns from animals was obtained from poultry; and poultry also provided an economic benefit to five other households in conjunction with contributions from other animal species. The returns from poultry were mainly in kind, representing the opportunity cost of the birds or eggs consumed by the family. Besides providing a source of valued food for the family, poultry act as a cash deposit since they are easily sold within the community. Families sell some of their chickens, turkeys or pigeons when they need small amounts of cash. Also, they are widely consumed in the months of June and July, at the end of school, when it is customary for families to offer a party to celebrate that children have concluded their school year. All the other species provide returns in cash through the sale of their produce or their work (for ploughing teams).

Equids provided an important cash return in six households in SPT, where they contributed to an average of 45% of the income derived from animals (range 4 –92%), mainly through renting out the ploughing teams; and in two farms in LCM where they provided an average of 17% of the economic returns from animals. The main income in cash from horses and mules comes from renting them out for ploughing and cultivating the land, and these are the figures quoted. No opportunity cost is incorporated in relation to the contribution that owning the ploughing team represents to those households who own large equids. In another study, Arriaga-Jordán and Velázquez-Beltrán (1999) estimated that owning a ploughing team in San Pablo Tlalchichilpa represented an opportunity value of Mex. 898/ ha (US$ 96 /ha), or Mex. 2,243. (US$ 241) per 2.5 ha farm when considering all the tilling activities plus the transport of the harvest. These figures should be included in the economic contribution of draught equids to the livelihoods of families who own them to have a more complete understanding of their role. The important role that work animals have in smallholder farming systems is recognised in many other areas where their use can result in improvements in food production and financial gains from the hiring out of work animals to non-owners (In Africa, for example, Muchena, 1989; Barrett, 1992; Shetto *et al.*, 2000).

In addition, there are a number of other important economic benefits from working equids that cannot be accounted for in money terms, such as the diverse functions they perform as pack animals. In this sense, equids, particularly donkeys, play an important role in carrying water, firewood, clothes, food and other goods for the *campesino* family. It is not common for families to rent in animals for these activities, and they usually are the responsibility of women and children. In this instance equids play an important role in reducing drudgery, as otherwise people would have to carry heavy burdens (Arriaga-Jordán and Velázquez-Beltrán, 1999). This is also true elsewhere. In Ethiopia for example pack donkeys make a large, and rarely quantified, contribution to the household activities and income through their transport activities (Pearson *et al.*, 2002).

In eight of the 25 households (five in SPT and three in LCM) sheep sales (or consumption) provided incomes ranging from 7 to 100% of total income from animals, representing 45% of income from animals in these farms with a mean return of Mex.1,540 /year (US$ 166/ year), representing 12% of the local minimum wage. Because of the high demand for sheep, they are also easily convertible assets, and given their larger size they act as medium-term savings. In villages like LCM, sheep are used to save money towards ownership of a ploughing team. Sheep are used to cover cash needs when needed, and if the family can afford it, they are sold to buy, or exchanged for, a bull calf to initiate the ploughing team. These families will pair up with other families in a similar situation to complete the pair of bulls. In time, and if no other need arises, they will sell or exchange their bull, and sheep if needed, to buy two bull calves and have their own pair of ploughing bulls (González-Díaz *et al.*, 1996). In SPT, where draught bulls have been replaced by large equids, sheep are sold to buy, or exchanged for, beef cattle, or more recently, dairy cattle. This is explained by the figures on the economic benefits brought about by cattle. Most sheep sales, as well as for other animals, take place around December and January, which also coincides with the dry season. Fifty eight per cent of sheep sales take place in this period.

Poultry are the most frequent animal species found in *campesino* households. Chickens, turkeys and to a lesser extent ducks represented benefits for 17 households, representing a mean income of US$ 58/ year, equivalent to 4% of the minimum wage in the area (Table 7.2). Seven households obtained 100% of the income in cash or kind from animals from poultry, where they brought in over US$ 100/year to each household, or 8% of the daily minimum wage per day. Although income from poultry is usually smaller than from other species, they represent a fundamental asset in *campesino* livelihoods, providing good quality food for the family, a source of cash from the sale of eggs or chickens, and they are often the only resource of women; so that projects on the development of the poultry component of smallholder systems have proven successful in alleviating poverty and improving nutrition (Dolberg, 2001).

The other species reported as contributing economically to households were rabbits; which represented the main source of income from animals in farm SPT7 that has the highest number of animal species (seven) and is the smallest of the monitored farms in San Pablo Tlalchichilpa (0.5 ha), so that diversification both in terms of plant and animal species is one of their productive strategies. The remaining 17% of the income from animals in this farm was obtained from selling pigeons, and this family also breeds canaries that are sold locally. No figures on the income from selling these ornate birds were provided.

An important contribution from cattle, equids and sheep is the manure, which *campesinos* value as an organic fertilizer, stating "the *maize grows better in land that has received livestock manure*". Work in East Africa has reported on the importance of manure in maintaining soil fertility and enhancing biodiversity (Kang'ara *et al.*, 2001); and Ayalew *et al.* (2001) ascribed an opportunity value for manure in terms of augmenting the chemical composition of soils as well as in improving its physical structure. However, it was not possible to establish a realistic opportunity value to the fertilizer contribution of manure, although some farmers say that up to half the chemical fertilizers needed for maize can be saved by applying cattle or sheep manure (and to a lesser extent equid manure).

The lack of economic contribution from animals in five farms in Mayorazgo was due to the small number of animals on these farms, and due to the fact also that most chickens died when affected by the disease outbreaks that periodically decimate *campesino* poultry populations in central Mexico. At the time of this report, all five households had bought new chicks to re-establish their flocks and benefit from chicken meat and egg production.

It is important to note in the analysis of the economic contribution of animals to *campesino* livelihoods the case of those farms that obtained returns from various animal species. This illustrates the importance of a diversity of animal species that enables a range of income from different sources and at different times during the year. The diversity of animal species also spreads risks of losses from disease, and given the different times when incomes are obtained (i.e. rent from ploughing equids is mainly between February and June), it enables the household to budget in the short, medium and long-term. An important aspect of the study is that the income obtained from the animal component of these *campesino* systems is independent from farm size. The R^2 for the linear regression of income generated on farm size was just 0.13 for SPT and 0.16 for LCM.

Ejido San Cristóbal

Under the more favourable conditions on the valley bottom, campesinos in Ejido San Cristóbal have incorporated dairy production into their farming systems, such that milk sales have become the main source of agricultural incomes, even to those farmers who identify themselves still as maize producers and whose herds are only seen as complementary (Espinoza *et al.*, 2000). Even under traditional feeding strategies based on maize straw and concentrates, milk production represented 64% of the margins per hectare of the traditional maize (straw) - cattle system (Arriaga-Jordán *et al.*, 1999a). Results showed that herds of five or more cows had profitable returns. When improved pastures and forages are included in the systems, important economic impacts are observed, since farmers following a feeding strategy based on grazing improved pastures, with maize silage during the dry season, and limited amounts of concentrates may obtain up to double the margins per hectare than under traditional feeding strategies (Arriaga-Jordan *et al.*, 2002).

Margins per day of family labour, which may be the most important economic variable for *campesino* farming families are around US$ 9.00 /day per person, with up to US$ 36.50/day obtained by a successful farmer. Considering that the opportunity costs for hired labour in the area is around US$ 4.50/day margins per day of family labour are on average two times higher in the dairy enterprises (Arriaga-Jordán *et al.*, 2002).

The demonstration module in the plot of the local school has provided a site where different trials have been undertaken to strengthen the above mentioned results without subjecting farmers themselves to what they perceived as risks for their systems. The module has also enabled access to all farmers from Ejido San Cristóbal or other neighbouring villages who can see and learn from what is being tried in the module. Table 7.3 shows the economic results obtained during the three years the module has been in operation (Arriaga-Jordán *et al.*, 2002b).

Table 7.3. Productivity indicators of the demonstration module for the development of appropriate feeding strategies in Ejido San Cristóbal (1999 – 2002).

	Years			
	1°	*2°*	*3°*	*Mean*
Sold milk/total land (l/ha)	7,782	12,047	11,929	10,586
Sold milk/pasture (l/ha)	12,971	16,063	19,881	16,305
Margin/pasture ha (US $/ha)	1,975	2,396	2,870	2,414
Margin/total ha (US $/ha)	1,185	1,438	1,722	1,448
Sold milk/cow (l/cow)	3,243	4,016	4,260	3,840
Margin/cow (US $/cow)	494	599	615	569
Cost of sold milk (US $/l)	0.18	0.16	0.19	0.18
Income/expenditure (US $)	1.84	1.91	1.64	1.80
Income/day (US $)	8	10	12	10

Mexican peso (Mex.) 9.60 / US Dollar ($)

Milk sold per cow has reached around 4,200 l/cow/year, with lactation yields of around 4,500 l/cow/year, with margins per hectare well above other agricultural activities. Basing production on forages and limited concentrate use, gives gross margins that would be equivalent to an average income of US$ 10.00 per day (Arriaga-Jordán *et al.*, 2002b).

This work in Ejido San Cristóbal has shown the potential for the productive use of the limited land resources of *campesino* farmers in the face of the difficult economic conditions of other agricultural activities, mainly maize grain production. Besides attractive incomes, small scale dairy production provides stability by the daily sale of milk, and enables *campesino* families to budget their expenses. These benefits and security brought about by the small dairy herds mean that family members do not need to migrate out of their villages (Espinoza *et al.*, 2000; Wiggins *et al.*, 2001).

Conclusions

This work provides evidence of the fundamental role that animals have in *campesino* livelihoods, since they make use and add value to resources which otherwise would or could not be used.

In the *Mazahua* systems, under unfavourable agricultural conditions, a diversity of animal species enables the utilization of different resources, such that competition is minimized among the different animal species. The two species that provide most economic benefit, in cash or in kind, are cattle and equids. Sheep provide the means for mid-term savings, which provide the resources to purchase cattle, the main providers of economic benefits. These aspects result in a more intensive use of the limited resources of *campesino* families, which provide incomes in cash, as well as the provision of good quality food, two aspects of maximal importance for *campesino* communities given the poverty that they face.

Animals also represent a source of savings (in many instances the only one accessible to *campesino* families) which provide them with money in times of need, be it illness or disease, or to meet social commitments like weddings, end of school year and family parties, funerals, or religious festivities. In addition, having work animals on the farm enables the cultivation of the land with less drudgery than if it had to be done by hand, and the role of donkeys for the domestic chores literally takes burdens off the shoulders of women, children and the elderly, who would otherwise have to do the carrying.

Under the more favourable conditions of the valley in Ejido San Cristóbal, the feasibility of small-scale dairy farming has proved successful in achieving goals for systems that generate income, food and occupation.

The integration of animals in the farming systems both in the Mazahua campesino systems on the hill-slopes, as well as on the campesino dairy production systems, improves their livelihoods. There is still work to do in terms of lowering the use of external inputs and on improving the environmental and social sustainability of these systems.

Acknowledgements

The authors express gratitude to the campesino families of the sector of "La Era" in the village of San Pablo Tlalchichilpa and in the village of La Concepción Mayorazgo in the municipality of San Felipe del Progreso, and in Ejido San Cristóbal, municipality of Almoloya de Juárez, for their participation in the projects, their kindness, hospitality and their disposition to share and exchange their knowledge and experience. Also we are grateful to Mr. Gabino Nava Bernal, Ms. Ana María Pedroza Fuentes, Mr. Leon Velázquez Beltrán, Mr. Benito Albarrán Portillo and Mr. Anastacio García Martínez for their support in the field work. The financial support of the United Nations University Project on "People, Land Management and Environmental Change" (PLEC) for the work in San Felipe del Progreso, and to the Consejo

Nacional de Ciencia y Tecnología (CONACYT) of the Mexican government for the work in Ejido San Crsitóbal, and of Universidad Autónoma del Estado de México, is gratefully acknowledged.

References

Arriaga-Jordán, C.M. (1996). Estrategias de Alimentación de Bovinos Lecheros en Sistemas de Producción en Pequeña Escala. [Feeding strategies for dairy cattle in small-scale production systems]. *Mejoramiento de los Sistemas de Producción de Leche en Pequeña Escala. [Improvement of small-scale dairy production systems]*. Edited by Castelán Ortega, O. Universidad Autónoma del Estado de México, Toluca, México, pp. 45-68.

Arriaga,-Jordán C.M., González-Diaz, J., González-Esquivel, C., Nava-Bernal, G. and Velásquez-Beltrán, L. (1997). Caracterización de los sistemas de producción campesinos en dos zonas del municipio de San Felipe del Progreso, México. Estrategias contrastantes. [Characterisation of *campesino* production systems in two areas of the municipality of San Felipe del Progreso, Mexico. Contrasting strategies]. *Investigación para el desarrollo rural: Diez años de experiencias del CICA. [Research for rural development. Ten years of experience at CICA]*. Edited by Rivera-Herrejón, G., Arellano-Hernández, A., González-Diaz, L. and Arriaga- Jordan, C.M. Coordinación General de Investigación y Estudios Avanzados. Universidad Autónoma del Estado de México, Toluca, México, pp. 171-197.

Arriaga-Jordán, C.M. and Velázquez-Beltrán, L. (1999). Economics of draught animal ownership in smallholder campesino (peasant) hillslope agricultural production systems in the State of Mexico. *Draught Animal News* 31:12-15.

Arriaga-Jordán, C.M., Espinoza-Ortega, A., Rojo-Guadarrama, H., Valdés-Martínez, J.L., Sánchez-Vera, E. and Wiggins, S. (1999a). Aspectos socio - económicos de la producción campesina de leche en el Valle de Toluca: I. Evaluación económica inicial. [Socio-economic aspects of *campesino* dairy production in the Valley of Toluca: I. Initial economic appraisal]. *Agrociencia* 33:483-491.

Arriaga-Jordán, C.M., Espinoza-Ortega, A., Rojo-Guadarrama, H., Valdés-Martínez, J.L., Sánchez-Vera, E. and Wiggins, S. (1999b). Aspectos socio - económicos de la producción campesina de leche en el Valle de Toluca: II. Aspectos sociales. [Socio-economic aspects of *campesino* dairy production in the Valley of Toluca: II. Social aspects]. *Agrociencia* 33:457-461.

Arriaga-Jordán, C.M., Flores-Gallegos, F.J., Peña-Carmona, G., Albarrán-Portillo, B., García-Martínez, A., Espinoza-Ortega, A., González-Esquivel, C.E. and Castelán-Ortega, O.A. (2001).

Participatory on-farm evaluation of the response to concentrate supplementation by cows in early lactation in smallholder peasant (*campesino*) dairy production systems in the highlands of central Mexico. *Journal of Agricultural Science,Cambridge* 137:97-103.

Arriaga-Jordán, C.M., Espinoza-Ortega, A., Albarrán-Portillo, B., García-Martínez, A. and Castelán-Ortega, O. (2002a). Desarrollo de estrategias de alimentación apropiadas para sistemas campesinos de producción de leche: una experiencia económicamente viable. [Development of appropriate feeding strategies for campesino dairy production systems: An economically viable experience]. *3ª. Reunión Nacional de Investigadores de la Problemática Económica Pecuaria.* Universidad Autónoma Chapingo, Chapingo, México.

Arriaga-Jordán, C.M., Albarrán-Portillo, B., Espinoza-Ortega, A., García-Martínez, A., and Castelán-Ortega, O.A. (2002b). On-farm comparison of feeding strategies based on forages for small-scale dairy production systems in the Highlands of Central Mexico. *Experimental Agriculture* 38:375-388.

Ayalew, W., King, J.M., Burns, E. and Rischkowsky, B. (2001). *Economic evaluation of smallholder subsistence livestock production: Lessons from an Ethiopian goat development project.* Nota di lavoro 107.2001. The Fondazione Eni Enrico Mattei Note di Lavoro Series Index: htpp://www.feem.it/web/activ/_activ.html

Barrett, J.C. (1992). *The economic role of cattle in communal farming systems in Zimbabwe.* Overseas Development Institute (ODI) Pastoral Development network. Network Paper 32b. March 1992.

Brookfield, H., Stocking, M. and Brookfield, M. (1999). Guidelines on agrodiversity in demonstration site areas. *PLEC News and Views (Special Issue on Methodology)* 13:17-31.

Chávez-Mejía, C., Nava-Bernal, G. and Arriaga-Jordán, C.M. (1998). Elements of biodiversity in homegardens in the highlands of Central Mexico. *PLEC News and Views* 11:18-19.

Dillon, J.L. and Hardaker, B.J. (1993). *Farm management research for small farmer development.* FAO Farm Systems Management Series. Food and Agriculture Organization of the United Nations, Rome:

Dolberg, F. (2001). A livestock development approach that contributes to poverty alleviation and widespread improvement of nutrition among the poor. *Livestock Research for Rural Development* 13 (5). http://www.cipav.org.co/lrrd/lrrd13/5/dolb135.htm

Espinoza-Ortega, A., Arriaga-Jordán., C., Ramírez-Miranda, C. and Wiggins, S. (2000). La Reconversión Productiva en el Estado de México: Los Productores del Valle de Toluca ¿Maiceros o Lecheros?. [Productive change in the State of Mexico: Farmers in the Valley of Toluca, maize or dairy farmers?] La Ganadería en México: Globalización, Políticas, Regiones y Transferencia de Tecnología. [Livestock production in Mexico: Globalisation, policies, regions and transfer of technology]. Edited by Cavallotti-Vázquez, B.A. and Palacio- Muñoz, V.H. Departamento de Zootecnia. Centro de Investigaciones Económicas, Sociales y Tecnológicas de la Agroindustria y la Agricultura Mundial (CIESTAAM). Universidad Autónoma de Chapingo. Chapingo, México, pp. 146-159.

González-Díaz, J.G., Arriaga-Jordán, C.M. and Sánchez-Vera, E. (1996). The role of cattle and sheep in campesino (peasant) production systems in the highlands of central Mexico. *Livestock farming systems: research, development socio-economics and the land manager.* Edited by Dent, J.B., McGregor, M.J. and Sibbald, A.R. Wageningen Pers, Wageningen, Neatherlands, pp.103-108.

INEGI (Instituto Nacional de Estadística, Geografía e Informática) (1996). Anuario Estadístico del Estado de México. [Statistical yearbook of the State of Mexico]. Edición 1996. INEGI, Aguascalientes, México.

Kang'ara, J.N., Ngoroi, E.H., Muturi, J.M., Amboga, S.A., Ngugi, F.K. and Mwangi, I. (2001). *The role of livestock in soil N fertility, biodiversity, land use, cultural and welfare change in Nduuri Embu, Kenya*. http://www.unu.edu/env/plec/culsters/Eastafrica/nov2001/Kang'ara.pdf.

Muchena, M. (1989). The effect of ox sharing arrangements on the supply and use of draught animals in the communal areas of Zimbabwe. *Proceedings of a workshop on the socio- economic dimensions of livestock production in the communal lands of Zimbabwe.* Edited by Cousins, B., pp. 253-275.

Pearson, R.A., Alemayehu, M., Tesfaye, A., Allan, E.F., Smith, D.G. and Asfaw, M. (2002). *Use and management of donkeys in peri-urban areas of Ethiopia.* Draught Animal Power Technical Report 5, EARO/CTVM, Centre for Tropical Veterinary Medicine, Easter Bush, Roslin, EH25 9RG, UK.

Powell, J.M., Pearson, R.A. and Hopkins, J.C. (1998). Impacts of livestock on crop production. In: *Food, lands and livelihoods - setting research agendas for animal science.* Edited by Gill, M., Smith, T., Pollott, G.E., Owen, E. and Lawrence, T.L.J. Occasional Publication No 21, British Society of Animal Science, pp. 53-66.

Shetto, R.M. Mkomwa, S. and Simalenga, T.E. (2000). Entrepreneurship in animal traction: empowering rural initiatives. In: *Empowering farmers with animal traction.* Edited by Kaumbutho, P., Pearson, R.A. and Simalenga, T.E. Proceedings of the workshop of the Animal Traction Network for Eastern and Southern Africa (ATNESA), 20-24 September 1999, Mpumalanga, South Africa, pp.194-200.

Wiggins, S., Tzintzun-Rascón, R., Ramírez-González, M., Ramírez-González, R., Ramírez-Valencia, F.J., Ortiz-Ortiz, G., Piña-Cárdenas, B., Aguilar-Barradas, U., Espinoza-Ortega, A., Pedraza-Fuentes, A.M., Rivera-Herrejón, G. and Arriaga-Jordán, C.M. (2001). *Costos y Retornos de la Producción de Leche en Pequeña Escala en la Zona Central de México. La lechería como empresa. [Costs and returns of small-scale dairy farming in Central Mexico. Dairying as an enterprise]*. Serie Cuadernos de Investigación. Cuarta Época 19. Universidad Autónoma del Estado de México, Toluca, México.

Woodgate, G.R. (1994). Local environmental knowledge, agricultural development and livelihood sustainability in Mexico. In: *Strategies for sustainable development.* Edited by Redclift, M. and Sage, C. John Wiley and Sons, Chichester, UK, pp.133-170.

8

Small animal species in the livelihoods of small-scale farmers in tropical Bolivia

R.T. Paterson[1] and F. Rojas[2]

[1]*Sustainable Agriculture Group, Natural Resources Institute, The University of Greenwich, Chatham Maritime, Kent ME4 4TB, UK. Current address: 13 Damer Gardens, Henley-on-Thames, Oxfordshire RG9 1HX, UK*

[2]*Centro de Investigación Agrícola Tropical (CIAT), Casilla 247, Santa Cruz, Bolivia*

Introduction

In the Bolivian Department of Santa Cruz, the Provinces of Sara and Ichilo lie some 100 km North-West of the city of Santa Cruz de la Sierra, where they occupy an area of about 21,000 km^2. Most of the region is a flat, alluvial plain, 350-450 m above sea level, with young soils prone to localized, seasonal waterlogging, although the land becomes undulating and rises to 800 m as it approaches the foothills of the Andes to the west. The soils are moderately fertile with pH values often in the range of 4.5 to 5.5. Rainfall increases from about 1400 mm in the east to over 1800 mm in the west, about three quarters of which falls between October and May. The natural vegetation is mostly tall, evergreen forest (Paterson *et al.*, 2001).

Some 55% of the total population of 90,000 people live in about 300 rural communities in groups of 25 to 150 families (Roca, 1998). The farm families come from two quite distinct ethnic groups of almost equal size. The local lowland people make up just over half of the present population and are descended from Spanish immigrants, usually with some degree of genetic influence from the original lowland Indian population. The other group (46% of the present day total) is composed of almost pure-blood Indians who have migrated into the area in recent decades, from the highland Departments of the country, mainly Cochabamba and Potosí. These ethnic origins are important in terms of attitudes, aspirations and lifestyles. Lowlanders tend to grow a wider range of subsistence crops than their neighbours, including native fruit trees and local vegetables. They are less market-orientated than the highlanders, favouring a more varied diet and a less hectic lifestyle over attempts to maximise their incomes (Román, 1998).

Within designated colonization areas, usually after removal of the best, high-value tropical hardwood timber, the state assigns blocks of 30-50 ha of forest land to individuals or families, for agricultural use. The recipients are usually extremely poor, with access only to a few hand tools. They initially clear small areas of land to produce subsistence crops (rice, maize, vegetables, etc.) under a slash-and-burn (swidden) farming system, rotating the cropping area around the farm when falling fertility or increasing weed problems lead them to abandon the original area back to bush fallow. The farmers generally lack the capital to work more than a few hectares of their land at any one time (Thiele, 1991). Some of the immigrants from the highlands have experience of animal traction in their areas of origin, but the heavy soils and the presence of roots and trunks make these techniques impracticable in the lowlands, even if enough capital were available to finance the use of oxen. The agricultural activities are largely concerned with subsistence farming, although excess production, destined for the city, is sold locally.

In 1992, it was estimated that 74% of the farm families in the region had total annual incomes below US$1100, with 25% of them receiving less than US$300 (Roca, 1998). Prices for farm produce have increased over the past decade, but casual labour is still paid at a rate of about US$5 per day and total family cash income is frequently in the range of US$1000 – 1200 per year (Román, 1998).

The heavy agricultural work on the farm (land clearance, land preparation) is usually undertaken by adult males, but in both ethnic groups, the whole family is involved in lighter field work (planting, weeding and harvesting). When present, cattle are generally managed by men, but milking of the cows and caring for small animal species are duties usually undertaken by the women and children. Small animal species appear on the farms of members of both ethnic groups at a very early stage in the development of the system. These animals generally contribute to food security, although, in times of need, excess production not required for home consumption, is sold. Scavenging, local breed chickens are present on over 90% of farms in the region (Román, 1998) and local ducks are common on farms where there are natural streams. Under traditional management, all poultry and most other small animal species scavenge for almost all of their feed, eating leaves, seeds, worms and insects. They receive very small amounts of supplements, usually cracked grain, surplus and over-ripe fruits and household scraps.

The lowlanders maintain a wider range of animal species than the immigrant highlanders, sometimes including native animals such as armadillos, which are captured in the surrounding forest and grown on to slaughter weight in captivity. Amongst their poultry, they often keep several varieties of chickens, together with ducks, geese and guinea-fowl. In contrast to this, the highlanders usually keep only chickens. The latter group do, however, have guinea pigs and tropical hair sheep, species which are seldom found on the farms of lowlanders. Previous livestock research and development work in the region concentrated on cattle, since the farmers of both ethnic groups all repeatedly expressed their desire to become involved in either dairy or beef production. It was noted, however, that the poorest members of society were usually unable to afford the investment needed to establish a herd of more than one or two cows. Because of their long gestation period and low calving rates associated with poor nutrition and the presence of tick-borne diseases, these animals made little contribution to family income and consumption (Breinholt, 1982). The programme of work discussed in the following pages was undertaken to define the role of small animal species in the households in the region; to identify the major production limitations; and to promote their use in the farming operations conducted by poor people on small-scale farms in the forest margins.

Materials and methods

The work described here took place in a fully participatory manner, where farm families (both adults and children of school age) and local extension staff worked in full collaboration with a multi-disciplinary team of researchers in both biological and social sciences. When the research started, there was an almost complete absence of technical information on the performance of small animal species and the problems facing them in the target area. A programme of many months work would be necessary to generate this vital information and for this reason, a conscious decision was taken to start the biological research first and to then follow this up with the socio-economic studies that would be necessary to complement the biophysical information.

Initial informal surveys in the target area identified chickens, ducks, pigs, hair sheep and guinea pigs as the most common livestock and subsequent work concentrated on these species. Monitoring of representative farms took place over a complete year (18 months for sheep), where all events (births, deaths, sickness, accidents, utilization, sales, feed offered, etc) were carefully recorded. This was done on 11 farms, located in the four distinct communities of Barrientos, San Rafael, San Miguelitos and Potrerito. Initially, technical staff visited each farm on a weekly basis to ensure that the records were kept up-to-date, but as the families became used to the routine, the period between visits was increased to two weeks. School-age children were often involved in the recording of the data, particularly where their parents were illiterate, or lacked confidence in their writing abilities.

As far as possible, technical interventions were avoided during this period, to establish the production patterns of the chosen species under traditional management. The major production problems were also determined, using, where appropriate, the assistance of a local diagnostic veterinary laboratory. The methodology employed during this phase of the work has been fully described by Paterson *et al.* (2001).

At the conclusion of this period, possible interventions were discussed in open meetings with the farm families and their neighbours, held in each of the four communities. At the conclusion of a presentation of the results obtained in the initial stages, the technical staff withdrew to allow the communities to freely discuss the findings and to choose which animal species and interventions would be tested in a subsequent programme of on-farm validation. The communities themselves nominated the individuals who would participate in this phase of the work, and neighbours were free to oversee the activities and to suggest modifications as they saw fit.

In parallel with the technical work on animal production, sociological and economic studies were also conducted in the same communities, using a range of participatory techniques, including semi-structured surveys, participatory rural appraisals, farm walks, maps, transects, resource flows and SWOT (strengths, weaknesses, opportunities, threats) analysis and wealth ranking (Román, 1999; Chamón *et al.*, 1999). This allowed an assessment of attitudes and aspirations in the communities, while measuring the contribution of traditionally managed small animal species to family livelihoods. The socio-economic methodology developed for this work was published for the benefit of other groups that may be interested in conducting similar work in other parts of the country or elsewhere (Chamón *et al.*, 2000).

Workshops and farm visits were arranged to publicise the findings and to confirm that the results, which had been obtained in a sample of communities, were applicable to the region as a whole. During the course of the work, results were published in a range of media, including technical fact-sheets, newspaper articles, booklets and scientific papers for submission to conferences and journals. At the conclusion of the work, mindful of the limited reading ability of many of the target farmers, CIAT (International Center for Tropical Agriculture) produced a series of three extension booklets on small animal production. These were based on illustrations and drawings, with a minimum of text, so that school-age children would be able to help their parents to understand the information presented. One booklet was dedicated to the prevention and control of common diseases (Choque *et al.*, 2002a), a second was on the construction of simple, rustic installations which would protect the animals from predators and the elements (Choque *et al.*, 2002b), while the third was on the use of commonly available feed resources for the animals (Lizárraga *et al.*, 2002). These have recently been translated into English and their publication will enable them to be used in other parts of the developing world with similar conditions and production systems.

Results

Average holdings of breeding animals varied widely between farms and over time, in response to family preferences and needs for cash and meat, but typically consisted of about 15 chickens, 6 ducks, 1 or 2 sows, up to 10 ewes and 15 guinea pigs. No single farm had all of these species, however. Chickens and pigs were kept by both highland immigrants and lowlanders, while ducks were only kept by the lowland people and sheep and guinea pigs only by the highlanders.

Attempts were made to identify cultural differences in the management practices used for small animal species on farms belonging to either highlanders or lowlanders (Rojas *et al.*, 2001). The study failed to define large differences, although, because of access to accumulated knowledge, the lowlanders made more use of traditional and household remedies to treat disease problems than did their neighbours. The highlanders had lived in a harsher environment in their original areas and possibly because of this, they were more inclined to provide rustic shelter for their animals. This was noted particularly with sheep and guinea pigs, species that are not normally kept by lowlanders. In the sections that follow, no attempt is made to distinguish between the ethnic groups in terms of the management used for their animals.

Under traditional management, poultry scavenge around the home compound for the bulk of their feed, receiving only occasional supplements of household scraps and cracked grain. Pigs also scavenge, while receiving occasional supplements of chopped cassava, etc., while sheep graze and browse along roadsides, in fallow-lands, or in small fields established for use by cattle. Guinea pigs are the exception, since they are usually kept in sheds, or in the family kitchen, where they receive their food as cut fodder and household scraps. None of the animals receive either vaccinations or veterinary treatment and most of them find their own shelter where they can.

As a result of the use of almost zero inputs, apart from the labour of women and children, which has a very low opportunity cost, any production in terms of eggs and meat from small animal species can be considered as profit, whether they are sold or eaten by the family.

Under these conditions, calculations based on a series of in-depth interviews and a sound understanding of the incomes, yields and budgets of representative families showed that the return from small animal species can represent up to 30% of annual family income, when home consumption is costed at the prevailing local market prices. This figure varied greatly between farms and even between seasons, since the holdings of small animal species were subject to large changes during the course of the study, but in view of their importance to family income and welfare, it was clear that attempts to increase their productivity could have a major effect on the small-farm sector.

Chickens

Chickens are kept by over 90% of farmers in the area, irrespective of their ethnic grouping. On most farms, eggs were collected for home consumption and occasional sale in almost all months of the year, with a peak in the period from August to November. Chicks hatched throughout the year, with greatest numbers from May to July. Overall, the annual productivity was 5.8 chicks per mature hen, although there was great variability between farms, due to the obvious trade-off between egg collection and the hatching of chicks. Typically, the annual sum of eggs collected and chicks observed (net egg production after losses to weather and predators) was in the range of 20-40 per breeding female. A number of native predators took both eggs and newly hatched chicks, but losses of older birds in this way were minimal. The heaviest mortality of chicks resulted from diseases such as Newcastle disease, fowl pox, infectious coryza, fowl cholera and pullorum disease (bacillary white diarrhoea), and the losses seemed to be aggravated by cold, wet weather. Both internal (roundworms) and external parasites (mites, mange) were common and although they did not normally lead directly to mortality, they had an effect on general health and welfare of the birds.

Few mature birds were sold, but young males and older females were slaughtered throughout the year for home consumption. In most cases, two to four birds were consumed each month but, occasionally, holiday periods or family celebrations could increase this number.

During the on-farm validation period of the work, farmers agreed to provide rustic housing for their poultry, made of poles and palm thatch cut from the farm, together with purchased chicken wire. The adults rapidly accustomed themselves to sleeping in the shelters and laying eggs in the nest boxes provided, although they continued to scavenge during the day for the bulk of their food. Vaccination was practised against Newcastle disease and fowl pox, while antibiotics and traditional remedies were used to treat diarrhoea when it occurred.

These measures increased the numbers of chicks observed and reduced losses, so that productivity per breeding female was increased by about 30 (Table 8.1).

Ducks

Ducks are not kept by immigrants from the highlands, because they are considered to be dirty birds that foul the home compound with their droppings. Under traditional management, no eggs were laid during the colder months (May to July) and females bred only once per year, with most eggs hatching in the period from October to March. On average, 6.7 ducklings hatched from each clutch of eggs, but more than half were lost to diseases, usually involving diarrhoea. No losses to predators were recorded during a monitoring period of a full year, although on one farm, two older birds died after being accidentally trampled by cows. There was no routine home consumption of either eggs or meat, but occasionally, a bird was slaughtered for a family celebration. Instead, there was a ready market for live or dressed birds, which often found their way, through intermediaries, to restaurants in the urban centres. Most birds were sold at a price set on sight, rather than by weight.

During the on-farm validation period, ducks were allowed entry into the shelters provided for the chickens and they received appropriate vaccinations and veterinary treatment with antibiotics as necessary. This resulted in a halving of the mortality rate of the ducklings and a doubling of the productivity of the species (see Table 8.1). During the course of the work, there was a growing trend towards setting the sale price by body weight, although this may have been a result of increasing sophistication in the market, rather than to any improvement in the levels of production.

One urban farmer, in a small town in the region, developed a semi-intensive, back-yard system of duck production, which produced excellent technical and financial results, even though it was based on purchased, rather than home-grown feed. Immediately after hatching, the ducklings were removed from the mother and kept inside the family house for two days, to avoid any possible adverse effects of cold and damp conditions. They were then placed outside in a small, sheltered enclosure of their own, where they were provided with clean water and a commercial poultry ration, together with vegetable scraps obtained cheaply from the local market. As they grew, they were removed to larger enclosures and in this way, they reached a live weight for slaughter of about 2 kg in a period of 3-4 months. They were then sold, by weight, to intermediaries, for the restaurant trade in the city. Under this system, diseases were not a problem, as there was no ready source of infection and any sick duckling was separated

Table 8.1. Annual productivity of small animal species.

Species	*Traditional management*			*Improved management*		
	Live young per breeding event	*Mortality before maturity (%)*	*Gross return per breeding female (US$/year*)*	*Mortality before maturity (%)*	*Gross return per breeding female (US$/year*)*	*Increase in productivity (%)*
Chickens	7.1	30.5	13.29	18.2	17.21	29.5
Ducks	6.7	54.1	12.05	27.4	23.82	97.7
Pigs	8.1	38.2	260.34	28.3	302.05	16.0
Hair-sheep	1.2	30.8	13.82	19.5	22.51	62.9
Guinea pigs	2.3	10.5	8.91	11.0	9.30	4.4

*Returns calculated at constant 1999 prices, US$1.00 = Bs5.50

from the rest as soon as symptoms were observed. All of the breeding females, kept with a drake in their own enclosure, laid two clutches per year and although fairly variable, losses of ducklings were kept to a maximum of about 10%. Annual gross returns per breeding female could be estimated at about US$48 after taking the costs of purchased feed into consideration. The owner considered that her flock of 8-10 breeding females provided more profit for much less effort than she could earn by the alternative of taking in laundry. This is a low-cost management system that could readily be copied by other families in the region. The profit margin could be increased by replacing the purchased feed with a largely home-grown a ration, based on commonly available feed resources such as maize, rice bran and a protein source, such as a meal made from soya or from the legume *Mucuna pruriens*.

Pigs

Many small farms run by both ethnic groups in the region kept one or two sows which scavenged for most of their food. They were free to roam around the home compound and nearby watercourses for most of the year, although they tended to be tethered close to the house when nearby annual crops were at a vulnerable stage of their development. They received no veterinary treatment and at best, only small amounts of chopped cassava tubers, chopped whole sugar cane and vegetable wastes to supplement what they could find for themselves. The fallen fruits from naturally occurring palm trees (*Attalea* and *Scheelea* species) formed an appreciable proportion of the diets of scavenging pigs, together with earthworms and roots obtained from low-lying areas or the banks of watercourses.

Only one or two farmers in each community kept a boar, which was readily lent to neighbours when required. Under these conditions, the average period between farrowings was calculated at 265 days. Litters were concentrated in the period from April to June, and again from September to November, with an average of 8.1 live piglets per litter, irrespective of the month of farrowing. Losses of piglets were generally high (up to 67%) and were attributed by farmers to accidents (crushing, drowning), cold weather and the lack of maternal milk. Research showed that where internal parasites were routinely controlled in mothers and offspring, losses over the first few months of life could be reduced to about 10%.

Young animals of up to 40 kg live weight were in great demand, particularly at times of public holidays (Carnival, Easter, Christmas) and for major family celebrations. Typically, of a litter of 6 pigs surviving to slaughter weight at from 4 to 6 months of age, one or two would be consumed by the family, but most would be sold locally, usually to intermediaries. Where sows farrowed twice in a year, an annual cash income of up to about US$400 could be derived, equivalent to the income from over three months of casual, daily paid work.

During the on-farm validation period, rustic enclosures were built, using poles and palm thatch, cut on the farm. A separate, shaded farrowing pen was included, to protect the newly born piglets from contact with other, larger animals. The pigs were vaccinated against classical swine fever and routinely treated to control both internal and external parasites. The result was to reduce the mortality of the piglets and to increase their growth rates (Table 8.1), although the improvements were limited by the unwillingness of the owners to provide adequate amounts of supplementary feed, particularly while the animals were confined in their pens. Although losses of piglets were reduced in the first two weeks of life, losses continued after this initial period as a result of accidents and an apparent lack of milk. More attention to the feeding of the dams in mid-lactation should allow a greater proportion of the litter to reach slaughter weight.

Hair sheep

Tropical hair sheep are kept only by highland immigrants, many of whom had a tradition of raising conventional, wool sheep in their original homelands. Lowland people have never been used to sheep, nor have they developed a taste for the meat, since wool sheep are poorly adapted to life in the humid tropics. For lowlanders, sheep are a totally new species and only a few families have recently started to express an interest in learning to care for them. Sheep are normally grazed and browsed on poor quality, volunteer pasture on fallow land and along roadsides, although occasionally, they are given access to better quality pastures which have normally been established for cattle. These planted pastures are usually sown to *Brachiaria decumbens*, *B. brizantha* or *B. humidicola*, although occasionally, areas are planted to a mixture of *Panicum maximum* with the pasture legumes *Macrotyloma axillare* cv. Archer, or *Pueraria phaseoloides* (tropical kudzu). They receive neither veterinary attention nor supplementary feeding, except for occasional access to a salt lick. Under this traditional management, where rams are constantly with the ewes, twinning is common, but mortality in the lambs prior to maturity is about 30% (Table 8.1). On four farms monitored over the course of 18 months, a total of 16 ewes lambed more than once. The average *inter partum* period was 251 days (range 144-424 days), including three ewes with more than a year between births. The average was 1.30 live young born per lambing.

The high level of lamb mortality was attributed by the farmers to accidents and lack of maternal milk, although technical staff were convinced that internal parasites were implicated in the problem. Diarrhoea also caused some losses, as did infections contracted by new-born lambs through an untreated navel. Some animals were slaughtered for consumption at times of family celebrations, although most were sold. Some farmers complained of difficulties of marketing animals for slaughter, because of the limited demand for mutton in the region, although others stated that there were no problems in selling young stock for breeding purposes.

Interventions included the provision of pens with a sheltered area to protect the animals at night from wind and rain, together with simple, veterinary treatments. These included the application of iodine to the navels of the new-born lambs, the treatment of diarrhoea with rehydrating salt (sugar and common salt in clean water), the use of anti-biotics to treat footrot and routine treatments to control internal and external parasites. These measures reduced the mortality of young animals to about 20%, with a consequent increase in productivity and profitability of the flock (see Table 8.1). There appeared to be a reduction in the average *inter partum* period since ewes seemed to come into season faster after lambing, but the validation period was too short to provide the data needed to confirm this suggestion.

Guinea pigs

Guinea pigs are traditionally kept in the Highlands of Bolivia and many immigrant families continue to raise these animals for home consumption in their new environment. Lowlanders have shown little interest in the species, because of their perceived similarities to rats. They are usually allowed to run free in the kitchen building of the house, although on some farms, they are confined in a separate, small shed. As well as receiving household scraps, they are also regularly given freshly-cut, or partially wilted forage, often of tropical kudzu, or of leaves of the mulberry tree (*Morus alba*).

When monitored, the average *inter partum* period for guinea pigs was 190 days, with an average of 2.28 live births per litter and mortality of 10.5%. No diseases were recorded and the low level of losses was attributed to accidents, sometimes involving dogs of neighbouring families. Theft of semi-mature animals was an occasional problem.

Suitable designs for pens were made available, which were raised to keep the animals about 100 cm above the ground. This provided them with ventilation, while protecting them from dogs and other potential predators. Internal divisions in the pen separated the sire from the growing animals and so reduced the danger of injury through

family and in the wider community. In immigrant families, the women tend to sell all farm produce and to administer the income generated, but amongst the lowlanders, the men generally take care of most sales and purchases. The traditional viewpoint in the latter group is that there is no income to be had from small animal species, so the men tend to ignore them, being content to leave their management to the women. At present, the sale of a small surplus provides the more marginalized members of the family with a modest income that they can dispose of as they see fit. It is not known if this will continue, but at the present time, it provides women and children with a measure of independence that has not previously existed.

Increased productivity from small animal species, which has previously been ignored by agricultural research and development bodies in the tropical regions of South America, has been shown to be easily achievable and this has attracted much attention from the authorities, both in the target area and elsewhere. To date, Belgian foreign aid has contracted the services of CIAT to conduct similar activities to improve productivity and to promote the use of small animal species in other parts of the province of Sara, while the Bolivian national government has provided funds for similar work in San Ignacio, in the eastern part of the country. The European Union is funding work in Warnes, near to Santa Cruz, while funding for the Vallegrande area, in the mid-Andean region to the west, has been requested. Interest has been expressed in other Bolivian Departments, such as Tarija and Cochabamba, together with the Salta region in the north of Argentina.

Conclusions

The implementation of simple, cheap, available recommendations for vaccination and parasite control and the provision of simple, rustic shelters, largely built from materials available on the farms, were shown to reduce losses of young animals and to increase the productivity and profitability of the holdings. This, in turn, increased the stability of the farming enterprises and reduced the need to hunt in the forest. In general, lowland families consumed more of the animal products, thus enjoying a better and more varied diet, while immigrant families tended to sell more produce to provide for an increased cash income. This was often treated as capital to be invested in other productive activities on the farm. Small animals are cared for almost exclusively by women and children and their social status was improved as a result of their increased contribution to family welfare.

The programme of research has attracted the attention of local and national government, leading to requests for promotional work with small animal species in a number of other provinces, both in Santa

Cruz and in the neighbouring departments of Cochabamba and Chuquisaca. The potential impact is, therefore, large, both in the direct target area and elsewhere in the region. The possibility of an international impact cannot be ignored, since similar conditions and problems occur in neighbouring countries, particularly in Brazil, Argentina and Peru, as well as in parts of Africa and the Indian sub-continent.

Acknowledgements

This paper is an output from a research project funded by the UK Department for International Development (DFID) for the benefit of developing countries. The views expressed are not necessarily those of DFID. [Project number R6774, Livestock Production Programme].

References

Breinholt, K. (1982). Annual milk yields and reproductive performance on small farms in the Bolivian tropics. *Tropical Animal Production* 7: 283-291.

Chamón, K., Joaquín, N. and Paterson, R. (1999). Sintesis y analisis de diagnósticos participativos con productores de pequeña escala en las provincias Sara e Ichilo. [Synthesis and analysis of participatory diagnoses with small-scale producers in the provinces of Sara and Ichilo]. Centro de Investigación Agrícola Tropical, Santa Cruz, Bolivia.

Chamón, K., Joaquín, N. and Paterson, R. (2000). Guía metodologica para la investigación con especies de animales menores en fincas de paqueños productores. [Methodological guide for the on-farm study of small animal species with small-scale producers]. Centro de Investigación Agrícola Tropical, Santa Cruz, Bolivia.

Choque, J.C., Rojas, F., Lizárraga, H., Fernández, W., Palomino, E. and Joaquín, N. (2002a). Recomendaciones para el control y prevención de enfermedades en aves, ovinos y cerdos. [Recommendations for the prevention and control of diseases in poultry, sheep and pigs]. Centro de Investigación Agrícola Tropical/Natural Resources Institute, Santa Cruz, Bolivia.

Choque, J.C., Rojas, F., Lizárraga, H., Palomino, E. and Fernández, W (2002b). Infraestructura rural básica para crianza de animales menores. [Basic rural infrastructure for raising small animals]. Centro de Investigación Agrícola Tropical/Natural Resources Institute, Santa Cruz, Bolivia.

Lizárraga, H., Rojas, F., Choque, J.C., Joaquín, N. and Fernández, W. (2002). Recomendaciones básicas para la alimentación de animales menores (aves, ovinos, cerdos y cuyes). [Basic recommendations for the feeding of small animals (poultry, sheep, pigs and guinea pigs)]. Centro de Investigación Agrícola Tropical/Natural Resources Institute, Santa Cruz, Bolivia.

Paterson, R.T., Joaquín, N., Chamón, K. and Palomino, E. (2001). The productivity of small animal species in small-scale mixed farming systems in subtropical Bolivia. *Tropical Animal Health and Production* 33: 1-14.

Roca, C. (1998). Contexto de la realidad de los sistemas de producción de los pequeños productores en la micro-región Ichilo-Sara. [The real context of production systems of small producers in the micro-region of Ichilo and Sara]. In: *Metodologías de Investigación Pecuaria en Sistemas de Producción de Pequeños Productores: Seminario Taller Internacional,* pp. 7-17. Centro de Investigación Agrícola Tropical, Santa Cruz, Bolivia.

Rojas, F., Chamón, K., Choque, J.C., Lizárraga, H., Fernández, W. and Palomino, E. (2001). Diferencias del sistema de manejo en especies de animales menores entre comunidades orientales y andinas de las provincias Sara e Ichilo. [Differences in the management systems for small animal species between lowland and highland communities in the provinces of Sara and Ichilo]. Centro de Investigación Agrícola Tropical, Santa Cruz, Bolivia.

Román, M.A. (1998). Influencia cultural en el desarrollo agropecuario de los pequeños productores de Santa Cruz. [Cultural influence in the agricultural development of small-scale farmers in Santa Cruz]. *Thesis, Universidad mayor de San Simón, Cochabamba, Bolivia.*

Román, M.A. (1999). Influencia cultural en el desarrollo socioeconomico de pequeños productores de las provincias Sara e Ichilo. [Cultural influence in the socio-economic development of small-scale farmers in Sara and Ichilo]. Centro de Investigación Agrícola Tropical, Santa Cruz, Bolivia.

Thiele, G. (1991). The barbecho crisis: revisited. Technical Report No. 1. Centro de Investigación Agrícola Tropical (CIAT)/British Tropical Agriculture Mission (BTAM), Santa Cruz, Bolivia.

9

Livestock, land use change, and environmental outcomes in the developing world

R.W. Blake[1] and C.F. Nicholson[2]
Departments of Animal Science[1] and Applied Economics and Management[2], Cornell University, Ithaca, NY 14853 USA

Demand drivers and linkages

Rapid predicted worldwide growth in demand for animal products to 2020, the so-called "next food revolution" in animal agriculture, portends complex interactions among people, biological and geophysical resources, and economic objectives. Consumer demand for beef and pork is expected to increase by 2.8% per year, somewhat slower than demand growth for poultry meat (3.1%/yr). Demand for dairy products is predicted to grow fastest, at 3.3%/yr (Delgado *et al.*, 1999). More people with greater *per capita* incomes, primarily urbanites in developing countries, are expected to purchase about 70% more meat and 90% more milk than they did in 2000. A restructuring of global food demands is expected: in contrast to current patterns, most (>60%) global production of meat and milk will be consumed by households in the developing countries (Cranfield *et al.*, 1998; Delgado *et al.*, 1999). This consumption shift will bring another major change: the livestock sector would become the *dominant value share* of global agricultural output. The key drivers of this change are income growth, population growth, urbanization, and increased opportunities for trade (Figure 9.1).

This future food demand scenario involves two elements of the so-called "critical triangle" of development (Vosti, 1995), the alleviation of poverty (by increasing food production and food security) and the economic growth to achieve it. Correspondingly, greater demand for products of animal origin signifies a critical development opportunity for producers, especially small-scale farmers in developing countries who rear the majority of the world's livestock. However, economic and poverty-alleviating objectives also confront competing goals and environmental costs that constitute the triangle's third element: factors that sustain, or that may threaten, the environment. For example, demand increases for meat from non-ruminants (poultry and swine) especially signify greater pollution from confinement systems and more land to produce feed grains, while greater demand for beef would stimulate more pasture use of land.

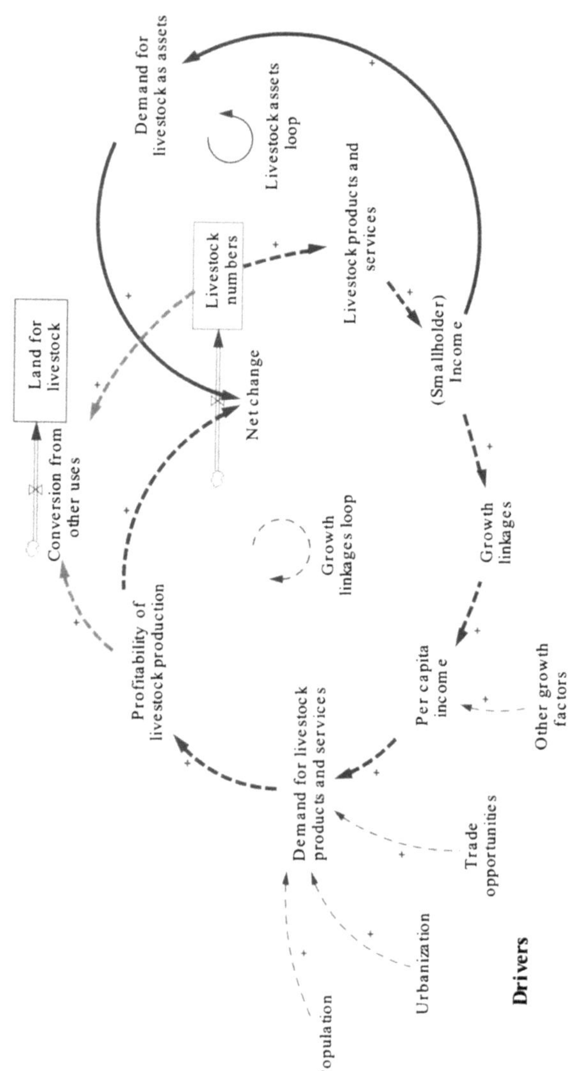

Figure 9.1. Key drivers of livestock product demand, stocks of land, and important feedback loops. The dashed growth linkages loop indicates that livestock ownership promotes income growth for smallholder farmers in the developing world. The solid livestock assets loop indicates that income growth can lead to the desire to own more livestock assets. The profitability of livestock and their numbers influence the conversion of land from other uses to livestock production (grey dashed arrows).

Selected potential impacts of increased demand for livestock products are illustrated in Figure 9.1. The key drivers result in additional demand for livestock products, which increases the profitability of livestock production. This, in turn, results in additional land being used for livestock and a larger global inventory of livestock. The development opportunities for smallholders are illustrated by the "growth linkages" loop, which relates an inventory with more animals to higher incomes for smallholders, whose spending of it generates additional income for other groups. In situations where the attractiveness of alternatives is low, increases in smallholder income can also be used to increase livestock numbers even further (the "livestock assets" loop). More livestock and land devoted to livestock production have potential negative environmental impacts through land conversion (e.g., forest clearing for pastures) and reductions in the productivity of land used for livestock, which results in the need for additional land clearing. Management strategies that conserve soil properties and support efficient nutrient cycling may offset degradation pathways, and increases in animal and land productivity may reduce the incentives for land conversion.

We identified some of the environmental risks, and recuperative effects, of animal agriculture in a recent article (Nicholson *et al.*, 2001) noting, however, that environmental conditions in developing countries will likely worsen before they *might* improve (Alexandratos, 1995, with world population growing daily by about 250,000 people, it may be unrealistic to expect improvement in environmental conditions). That article focused on the impacts of forest conversion to pasture in Latin America, biological diversity in East Africa, and global greenhouse gas emissions from livestock production. This presentation complements that previous one, focusing more broadly on the ecosystem impacts of conversion of land to agricultural uses. We concentrate on systems with ruminant livestock, discuss the linkages to growth in animal products demand, and identify appropriate policy and research.

A key question is "How can (should) international agricultural researchers, the development community, and policy makers support the beneficial aspects of the growth in demand for animal products while minimizing negative environmental outcomes?" The productivity of land and livestock must be improved in an economically and environmentally reasonable way to feed a human population whose growth is a primary force threatening Earth systems. Achieving these goals involves tradeoffs because win-win outcomes, where the ecosystem impacts accompanying livestock-related household activities are inherently neutral or favourable, may be realistically few. Thus, a second key question is "Who should pay, directly or indirectly, for changes in agricultural production practices to reduce or mitigate environmental impacts?" Furthermore, the characteristics of specific situations vary greatly, implying the need for site-specificity in research and policy initiatives. A case in point: whereas pollution from excess nutrients (and microbes) from animal waste is a problem in developed countries, many are the cases in the developing world where too few nutrients, not too many, constrain agricultural sustainability. For example, reducing the P content of excreta from swine with dietary phytase helps control P transfers to the environment (e.g., pollution). However, in tropical locations where manure is fertilizer, swine manure with normal high P content is premium fertilizer, especially for low-P soils or crops with a high requirement. Hence, specific cases and system complexity demand specific information, research knowledge, and policy interventions. General recommendations, therefore, may be inappropriate or ineffective. Also, agricultural research and development policies alone may be insufficient to resolve these problems if dominant "root causes" lie elsewhere. For example, population growth and increases in *per capita* resource use may prove to be far more significant an environmental problem than land conversion from livestock (Meadows, 2002). However, projections of population growth and assessments of its impacts differ. Therefore, we focus on examination of *current evidence* about impacts of livestock in the shorter run, without endorsing a particular view about the future in the longer term (50-100 yr).

The remainder of this chapter consists of three sections. The first section summarizes the environmental and ecological impacts of land conversion, emphasizing change from natural habitats to agricultural use (but also notes that intensification of existing agricultural systems involves similar effects and mechanisms). The second section discusses the environmental impacts and tradeoffs for the case of expanded beef production in the western Brazilian Amazon. This case demonstrates the desirability of site-specificity for research and policy design and the potential usefulness of systems modelling for *ex ante* evaluation of alternative interventions. The final section concludes with an overview of research and policy recommendations, noting the potential complexity of using agricultural technologies to reduce environmental impacts of livestock production.

Ecological and environmental impacts of land conversion

Relationships stemming from land conversion

Increased livestock production has a number of potential negative environmental consequences. Among these, the conversion of land from natural habitats to agricultural uses more generally, particularly in tropical forest areas, has been a key cause for concern for the past few decades. The key relationships among livestock production, land use change, and environmental outcomes are shown in Figure 9.2. Land use conversion from forest to pasture is driven by incentives for livestock production (which are increased by livestock demand), limited alternatives to livestock production in these areas, and perhaps more importantly, economic inequality and poverty that create incentives for forest clearing by migrants. Land is also cleared for cropping, and conversion of land from pasture to crops (and vice versa; Nicholson *et al.*, 1995). Clearing forest land reduces biodiversity and ecosystem services (e.g., pollination, pest control, flood control and water release). Current pasture management practices often

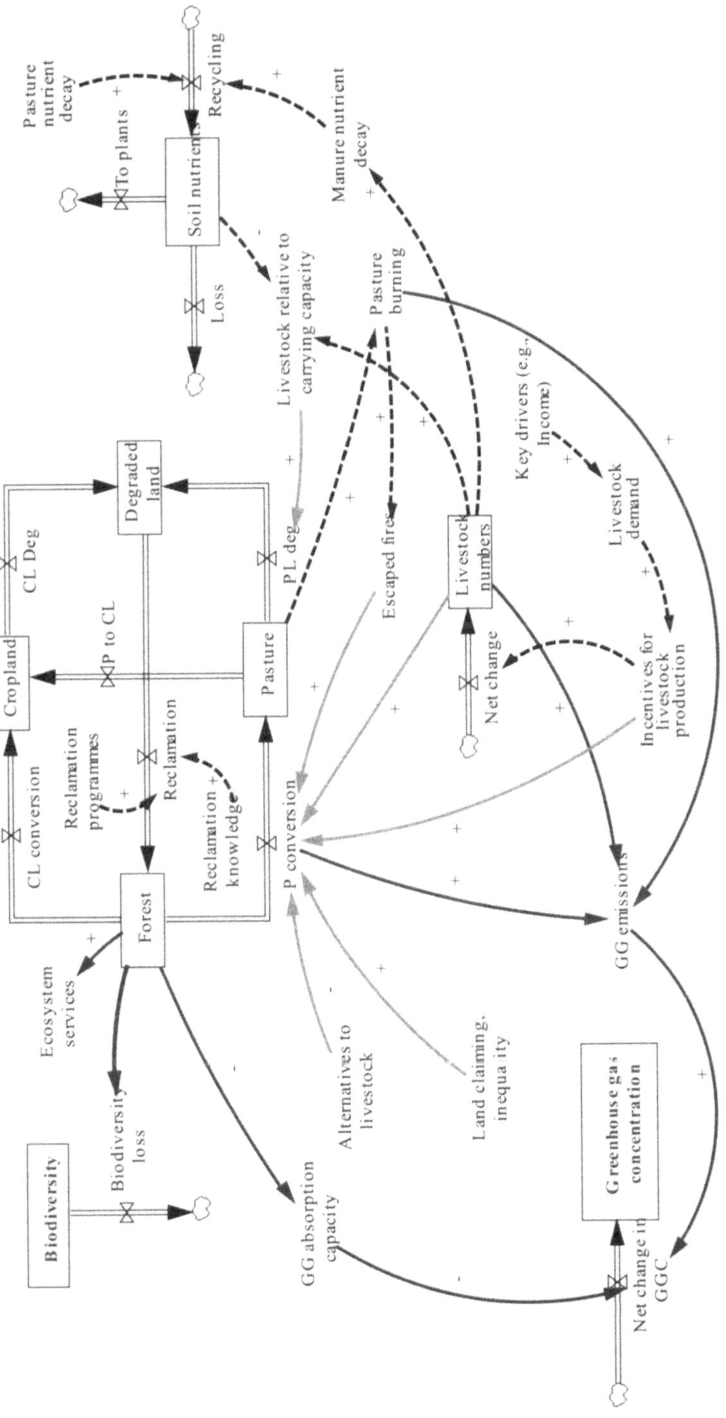

Figure 9.2. Relationships among livestock, land use change, and environmental outcomes. Dark grey arrows indicate key environmental or ecological impacts. Light grey arrows indicate effects on conversion of land to pasture. Dashed arrows indicate effects related to soil nutrient cycling. The polarity of individual effects is indicated as positive (+) or negative (-). [CL=crop land; GG=greenhouse gases; GGC=greenhouse gas concentration; P=pasture; PL=pasture land].

result in land degradation, which results in the need for more forest clearing, and pasture burning to promote regrowth and nutrient cycling leads to additional forest losses from accidental fires. The connection between land use change and greenhouse gas emissions is also indicated. Forest clearing (often through burning) releases large amounts of CO_2 to the atmosphere, and greater livestock numbers also increase greenhouse gas emissions. At the same time, the ability of terrestrial ecosystems to absorb gases (especially CO_2) is reduced by forest clearing.

The ecological literature provides quantitative estimates of the magnitude of some of these impacts and their importance. According to some observers, human demands on nature already may have exceeded regenerative capacity of the biosphere, which may bring an end to rapid agricultural expansion within 50 years (Tilman *et al.*, 2001a; Wackernagel *et al.*, 2002). "Just as demand for energy is the major cause of increasing atmospheric greenhouse gases, demand for agricultural products may be the major driver of future non-climatic global change" (Tilman *et al.*, 2001a). Various forecasts, differing in degree but not in direction (Laurance, 2001; Laurance *et al.*, 2001; Tilman *et al.*, 2001a; Vitousek *et al.*, 1997), anticipate important losses of habitat, biodiversity and ecosystem services by transforming more land to feed humankind, especially in Latin America and sub-Saharan Africa. These environmental modifications, involving greater inputs of N, P, water and pesticides, can affect economic returns in a variety of agricultural systems, including high input-output animal monoculture, integrated crop-livestock systems (e.g., smallholders), and modest-performance extensive systems (e.g., grazing). However, the extent of these economic effects in both the long term and the short term often is not well documented or acknowledged. General impacts of land conversion described in the literature include nutrient loading, agrochemical pollution, acidification, salinisation, eutrophication of surface water, emissions of greenhouse gases, and irreversible species extinctions. For livestock production systems in the developing world, greenhouse gas emissions, nutrient losses to the environment, and species extinctions are of greatest importance. Furthermore, substituting pastures or crops for forest cover increases albedo (the ratio of reflected to incident light), which may reduce precipitation and increase temperatures (Vitousek *et al.*, 1997). The remainder of this section focuses on the relationships between land use, habitat loss and fragmentation and biodiversity losses, and agroecological options for producing food and ameliorating the long-term environmental costs of doing so.

Habitat loss and fragmentation

Simplifying natural ecosystems (and biogeochemical cycles of C, N and P) with agricultural uses initiates processes that lead to biomass losses and species losses and substitutions: The abundance of biodiversity is directly related to the abundance of biomass. We noted in our previous paper that relationships between forest conversion and biodiversity have often been oversimplified, and that the dynamics between pasture, trees and forest remnants are complex. However, conversion of forest to pastureland is an important source of biodiversity loss from changes in regional climate, nutrient dynamics and biotic exchange, which implies the food revolution in animal agriculture could increase these losses. Burning is a principal pathway for initial forest clearing that releases nutrients stored in above-ground biomass. However, routine pasture management in many parts of Latin America involves periodic burning to promote pasture regrowth, control weeds, and release nutrients from senescent plant biomass. Unfortunately, about 40% of these pasture and slash fires escape to adjacent areas (Kauffman *et al.*, 1998). Burning practices resulted in about 17 million hectares of uncontrolled fires in Indonesia and Latin America in 1998. The estimated $19 to $25 billion cost of these fires (Cochrane, 2001), presumably mostly in sacrificed forest products, may greatly undervalue total losses also in habitat, species and atmospheric load. More light through more penetrable burnt forest canopy causes warming, which increases water losses and makes forests more fire-prone. In addition to destruction of habitat, burning pastures releases large quantities of C to the atmosphere. Therefore, pasture management in which less land is burned less frequently would protect forests and the services they provide.

Lost and fragmented habitats make it more difficult for plants and animals to meet needs and survive environmental vagaries (Laurance, 2001; Sala *et al.*, 2000). Although pollen flow could partly compensate fragmentation losses in certain cases, it requires that welfare and behavioural (spatial) responses of pollinators are assured (White *et al.*, 2002). Also, fragmented forest is more susceptible to fire from 'vegetation breeze', where clearings large and small draw moisture from the forest, substituting it with dry air. Dried forest edges are more likely to ignite when nearby pastures are burned (Laurance *et al.*, 2002).

Unfortunately, besides altered land use "the strength of interactions among drivers in their effects on biodiversity is *virtually unknown*" (Sala *et al.*, 2000). These interactions are poorly understood in low as well as high rainfall ecosystems. Zonal effects of cattlepost systems in the semi-arid Kalahari (Botswana) produce an environmental tradeoff between vegetation sacrificed and the amount of rangeland biodiversity that is retained (Perkins, 1996). Supported by borehole water, these dynamic systems rely on herd mobility and flexibility in a manner similar to wildlife. Water availability increases the area that can be used by cattle but results in severe land degradation due to excessive nutrient loading within 50 m of watering locations. This modest sacrifice in habitat permits cattlepost managers to make productive use of the remaining rangeland without significant degradation of that resource, which is now known to respond favourably (e.g., greater plant survival and productivity) to continuous grazing pressure and unfavourably to the lack of it (Oba *et al.*, 2000). In fact, land productivity of the cattlepost system exceeded that from subsidized commercial beef systems when all herd outputs (milk, beef, and draft power) were accounted (Perkins, 1996).

Cascading effects on species shifts and biodiversity losses

Transforming lands with marginal fertility to agriculture can have a large global effect because these locations sustain ecosystems of high diversity. About 18% of mammals, 11% of birds and 8% of plant species may currently be at risk of extinction from land conversion (Vitousek *et al.*, 1997), which implies significant agroecosystem losses such as wild genes for resistance to plant pests and disease, pollination by birds, insects, bats and other mammals (from neighbouring ecosystems), and pest outbreak control by predators (Tilman, 1999). Forest disturbance is one pathway that adversely affects understory insectivores, especially birds with limited dispersal ability whose vulnerability is heightened to forest edge effects, isolation and random events (Sekerciouglu *et al.*, 2002). Species least affected are those able to utilize clearings. Biological invasions also reduce species diversity and increase pest losses in crops and livestock (Pimentel *et al.*, 2000). Scarce soil nutrients are more completely utilised in high diversity systems, which constitutes a barrier to invading species (Tilman, 1999). Conversely, unconsumed nutrients may favour pathogens and pests, which may be a consequence of indiscriminate or uncalibrated fertilization of food and forage crops. For example, high soluble N in fertilized plants can invite larger populations of sap-feeding insects (Matson *et al.*, 1997), a phenomenon that may also apply to forage grasses. Differing from past recommendations relying on more fertilizer to produce more rice, today growers are sometimes urged to moderate their N use to reduce risk of yield losses from pest outbreaks. High external nutrient inputs on arable lands in rainy climates and erosion losses in overstocked arid lands also lead to less diversity. Increases in farm land have shrunk habitats, populations of non-domesticated animal species and their range of dispersal. This process imposes selection pressures against large species with low dispersal potential, which increases threats (e.g., weeds, pathogens and competitive microbes) to the environment and to agriculture (Western, 2001).

Species extinction is partly governed by the interaction between dispersal ability, habitat loss and climate change. Greater rainfall variability affects habitat and climate change and has resulted in population fluctuations and extinction of butterflies (McLaughlin *et al.*, 2002). Possibly exacerbated by substitution with grasses, larvae mortality rates increased from shorter overlap with the required plants. Environmental bottlenecks also invoke biological mechanisms of species loss. Predisposition to extinction is increased by inbreeding when the effective population size becomes small and insular, which results in losses of alleles with large effects on fitness and survival (Keller and Waller, 2002). Inbreeding depresses birth rate, birth weight, offspring survival, and resistance to disease and other stressors (e.g., predators), which makes individuals more likely to perish from vagaries of the environment. Less heterozygous sheep had more parasitism and were less likely to survive untreated helminthic infestation than more heterozygous individuals (Keller and Waller, 2002), an important consequence for producers who cannot afford chemical control.

Mitigating biodiversity losses requires clear understanding of the mechanisms in each case. On a global scale, this means a set of management portfolios matched efficaciously to biological and socio-economic needs by eco-region (Sala *et al.*, 2000). Productivity and diversity may be enhanced by complementarities and net favourable interactions among species, including resistance to pathogens and invaders, especially where nutrients are scarce (Western, 2001). Cropping with a species mixture may be useful (Tilman, 1999). This strategy is used by crop-livestock farmers in Ethiopia's low-rainfall Harar highlands (Kassa *et al.*, 2002), who densely sow maize and sorghum and selectively thin maize (sorghum) for forage if rainfall is low (high). Species complementarity and niche differentiation, for example from a mixture of forage genotypes, could exploit warm/cool and rainy/dry seasons, and deep/shallow soil nutrient profiles.

Agroecological options: given where we are, what can we do?

In summary, an important agroecological goal in the long-term is to assure sufficient land and resources for survival of most species (Vitousek *et al.*, 1997). This requires learning how ecosystems interact with, and recover from, environmental disturbances, including agriculture. In addition to converting less land to agriculture, it also means curtailing the cascade of interactions on habitats, species extinctions and complementarity losses, lost ecosystem services, and nutrient stocks (e.g., for food production, diversity, barriers to pests and pathogens, waste control). Western (2001) offered useful principles (Table 9.1) for reducing these environmental costs. Based on this guidance, a nutrient management research portfolio should address: 1) effective nutrient cycling between soils, plants and animals; 2) improved nutrient use efficiencies of plants and animals; 3) improved management of nutrient stocks and soil fertility; and 4) alternative uses of grazing land, especially in a more diversified agriculture that meets economic needs of producers. Other considerations are to recruit ecosystem services from other species by bridging habitat fragments with multi-use land buffers and corridors (e.g., along fence lines on pasture landscape), to plan agriculture with less land fragmentation and more ecosystem types with biodiversity hot spots preserved, the judicious use of agrochemicals and manure, fallow management (e.g., cover crops), development of productive crops with lower requirements for water, nutrients and pesticides, and integrated pest and pathogen management. However, achieving outcomes that are more environmentally sound often will require economic incentives, especially for small- and medium-sized farmers who are our key environmental stewards (Pinstrup-Andersen, 2001). The need for economic incentives raises the question of the political feasibility for successful implementation of these strategies on a sufficiently large scale to have notable impact.

Table 9.1. Principles for conserving ecosystem processes (from Western, 2001).

Maintain or multiply	*Minimize*	*Mimic*
Species richness, structural symmetry, key species/ functional groups	Erosion, nutrient losses and pollution emissions (e.g., fire)	Natural processes in production cycles
Internal regulatory processes and interactions (e.g., predator-prey)	Landscape simplification	
External diversifying forces	Landscape homogenisation	
Large habitat areas and spatial linkages between ecozones		
Ecological gradients and ecozones		

Pasture-based cattle production in Brazil's *arc of deforestation*

The foregoing discussion of impacts of land use change and potential actions is useful, but can be complemented by a discussion of impacts and responses for a specific area. Land use change and policy in the Brazilian Amazon, a topic of current debate (Laurance *et al.*, 2001; 2002), has been of great concern for at least the past two decades, and illustrates many of the general concepts identified above. Thus, we use results from studies of dual-purpose (milk-beef) and beef cattle systems in the Brazilian Amazon to discuss the potentials and limitations of modifying livestock production systems to achieve more desirable environmental outcomes.

The Amazon region of Brazil is essential to global biodiversity, to climate and regional hydrology, and to C stores by harbouring about 40% of the world's tropical rainforest. This region is also home to about one-fourth of the national herd of about 170 million cattle, which is currently reared extensively on approximately 43 million hectares. Forest losses are especially high in the four-state "arc of deforestation", including the westernmost Acre State, and could reach 5000 km^2/yr (Cochrane, 2001). This transformation is driven by population growth, logging and mining, spatial effects of development (e.g., roadways), and fire. Current trends suggest more habitat fragmentation, less rainfall from less evapo-transpiration and sequestered atmospheric moisture from smoke, higher land surface temperatures, species losses, and heightened risk of forest fires abetted by the drying effect of vegetation breeze (Laurance *et al.*, 2001; 2002). Although widespread pasture degradation is well documented in the eastern Amazon, the farm management practices (e.g., stocking rates, soil-plant nutrient relationships, burning frequencies) causing it have not been adequately studied. Thus, it is currently unknown to what extent alternative management would control degradation losses (and, therefore, reduce pressure for additional forest clearing).

To reduce environmental losses, Laurance and co-workers (2001) favour farm livelihoods earned from agroforestry and perennial cropping over "fire-maintained cattle pastures and slash-and-burn farming." Of course, these alternatives need to be economically attractive to be adopted, and the available evidence suggests limited potential (Vosti and Valentim, 1998). Alternatively, more intensive cattle production has been proposed to improve economic returns to labour and land while reducing deforestation pressure. However, greater profits from more intensive management (e.g., more beef and milk from more dietary inputs and labour) may provide incentives for additional land clearing. Pasture and cattle activities in Acre were estimated to be more profitable than other alternatives, which implied economic incentives to clear more forest (Carpentier *et al.*, 2000). Clearing was influenced by labour supply and access to markets.

Recent studies explored soil-plant-animal nutrient relationships, milk and beef potentials, and economic constraints of pasture-based cattle systems in Acre (the Western Amazon), where large fire-managed paddocks are lightly stocked with animals not differentiated by nutritional requirements. Findings indicated relatively low nutrient exports in animal products of 6 kg N/ha and 3 kg P/ha (compared to rice, beans or coffee crops) from farms with two 450-kg animal units per hectare (Rueda *et al.*, 2004). In contrast to studies in the Eastern Amazon, pasture productivity has been maintained for two decades. This relative longevity may result from better soil characteristics than elsewhere in Amazonia, but also from low stocking, which results in more-than-adequate pasture biomass availability, substantial nutrients recycled from leaf litter and less soil compaction. This evidence supports the idea that pasture management may play a key role in system sustainability. It also suggests there is potential for positive outcomes from research on alternative practices (and, in some cases, reclamation of degraded lands) for other parts of Amazonia.

The current cattle production system in the Western Amazon appears sustainable. However, improved access to beef markets in Lima, Peru and the Pacific region (through a recently completed highway) could create incentives for increased beef production in the region, implying that more intensive production systems may be needed to limit the ultimate extent of additional forest clearing. Substantial productive potential was identified for application of external nutrients to pasture to increase stocking rates and offtake. With herd nutrition based on accurate predictions of nutrients supplied by forages and required by animals (Tedeschi *et al.*, 2002), this translated to greater net economic returns from labour-intensive (i.e., potentially land saving) technologies to produce beef (but was not profitable for milk), and not by improving individual performance by better supplementation of diets (Rueda *et al.*, 2003). Although more labour would be required, beef production through judicious fertilization of grass-legume pastures and higher stocking rates may improve farm incomes and slow the rate of forest clearing.

However, the long-term dynamics of this intensification strategy are not known. Higher stocking rates imply shifts in the pools of nutrients to recycle from plant litter to readily decomposable animal excreta (Figure 9.3). To explore the potential impacts, a systems modelling approach can be useful. In a conceptual systems model, forage productivity and residual environmental health are represented as feedbacks among stocks and flows of four nutrient sources, soil, available pasture (forage) biomass, decaying pasture biomass, and decaying manure. The grazing-manure management pathway is especially important to herd productivity, external nutrient demand, and productivity and pasture health outcomes. Proper land allocations of manure through herd grazing management can reduce fertilizer requirements and maintain soil nutrient stocks, which supply plant requirements for growth. A key research question is the extent to which external nutrient inputs (e.g., mineral fertilizers) are required to sustain this system. Better knowledge of the sizes of the various nutrient pools and the rates affecting them (availability) would provide

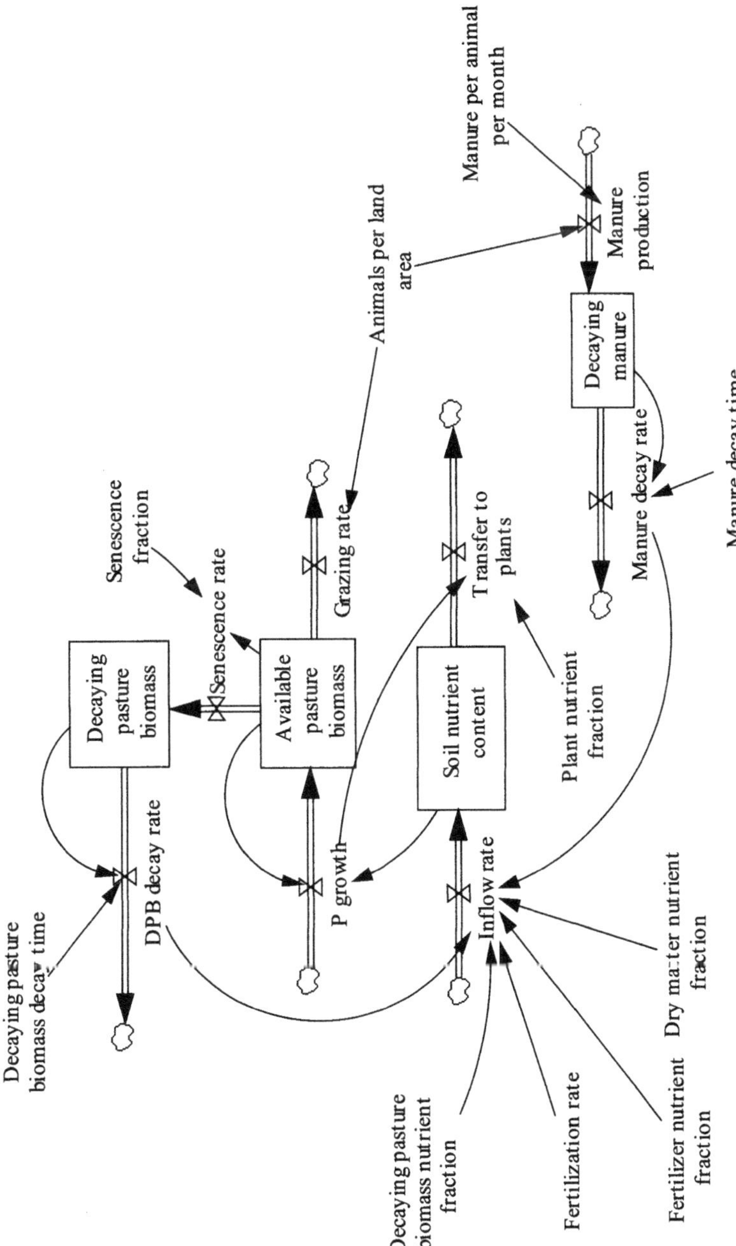

Figure 9.3.
Nutrient dynamics in the pasture-based cattle system of the Western Amazon of Brazil showing key nutrient pools and linkages between them. An increase in animals per land area increases the grazing rate, thus decreasing decaying pasture biomass with greater nutrient cycling through manure. The need for external nutrient inputs and system sustainability depend on the initial nutrient stocks and flow rates.

information crucial to more intensive management of the Acre production systems. However, it is again worth noting that beef demand is only one of several factors affecting land transformation and alternative agricultural technologies are unlikely to fully address its root causes (Nicholson *et al.*, 1995; 2001).

Although additional studies to better understand the dynamics of nutrient stocks in the Western Amazon can facilitate better responses to increased market access for beef production, available information suggests that certain strategies may help reduce negative environmental impacts. The recommendations in Table 9.2 emphasize opportunities for better integrated nutrient management under current economic conditions.

Table 9.2. Agro-ecosystem management opportunities for cattle systems in Acre, Brazil.

Pastures and animals	*Biodiversity*
Harvest pasture nutrients by intensive grazing; avoid burning	Design and build through pasturelands a bridge network of fragmentation-ameliorating, diversified forest blocks and corridors with income-earning potential (e.g., natural barrier fences) to preserve wildlife and re-generate ecosystem services
Herd nutrition based on accurate predictions of forage quality and animal nutrient requirements	
Group animals in paddocks according to their nutrient requirements	
Titrate stocking and herd size and provide drinking water to assure effective distribution of excreta across the landscape (to promote widespread nutrient recycling)	
Fertilize in accordance with the nutrients extracted and recycled from excreta	
Evaluate and utilize multi-species portfolios of grasses and legumes	
Focus management on herd reproduction and herd performance, not individual performance	

Limitations of the technology toolbox

To what extent can the development of agricultural technologies for cattle production systems in the Amazon region reduce the pressure for additional forest clearing? The multiple causes of forest clearing have already been mentioned, and these may, in fact, dominate attempts to increase (or maintain) the productivity of land and livestock. However, a more subtle challenge is that technologies that increase productivity may actually increase incentives to clear land. As Rueda and co-workers (2003) noted, profitable technologies that increase or maintain productivity of beef can reduce the need for land. However, if technologies are sufficiently profitable, they may result in incentives for additional forest clearing, so as to apply the new (more profitable) technology to a broader land base. The nature of this challenge is illustrated in Figure 9.4. Technology use depends on a number of factors, including the expected economic returns from doing so. If the technology increases the profitability of beef production, it creates incentives to clear additional land. If the technology is labour-intensive, and labour (for hire) is limited in the region, this may limit the impact of increased profitability on forest clearing (Angelsen and Kaimowitz, 2001). In the long-run, however, if the new technology is profitable enough, wages can rise and induce greater migration to the area, which will increase the availability of labour and, therefore, forest clearing. These basic principles apply broadly to many technological options, including those cited above. The basic result is that developing livestock production technologies that are profitable in the short run and do not exacerbate land use change is a major challenge.

Information needs and recommendations

More understanding is needed about the interaction of agricultural practices with environmental and ecological outcomes to develop practices that avoid further deterioration and restore accumulated damages. Models predicting habitat loss from conversion to agriculture and land degradation indicate that "a pool of species will eventually become extinct unless the habitat is repaired or restored" (Dobson *et al.*, 1997). Studying restoration processes of degraded land would build understanding about ecological communities and agro-ecological function. Priority sites include ecosystems where agriculture was short-lived, such as forests cleared in the Eastern Amazon. Integrated agro-ecological strategies include improved nutrient management, illustrated above for the Western Amazon. Complementary strategies include bridging (restoring) forest fragments with hospitable fence line corridors and windbreaks of native species to facilitate wildlife dispersals and ecosystem services (Sekercioglu *et al.*, 2002; Tilman *et al.*, 2001b). Investments are needed in productive and environmentally friendly technologies, especially for small- and medium-sized farmers who are equally efficient as large-scale operators and wise stewards of the environment. The technology toolbox should include new innovations based on external inputs and

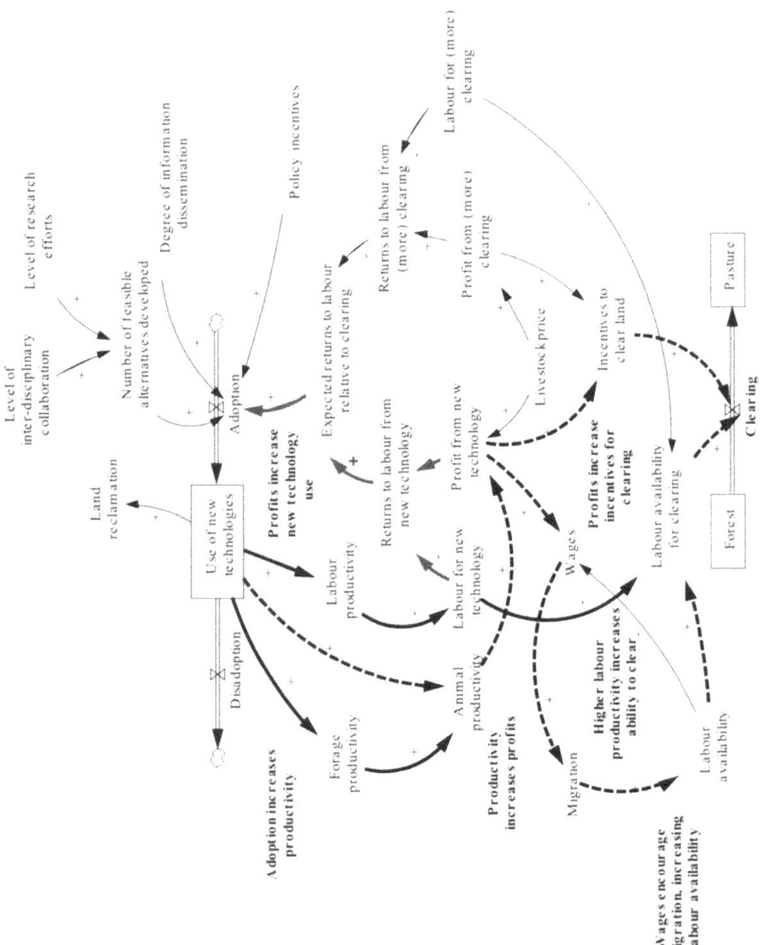

Figure 9.4. Relationships between technology adoption and forest clearing. Dashed pathways indicate key relationships between profitability and clearing. Heavy solid pathways indicate the relationships between productivities of land, labor and animals and clearing. Grey pathways indicate feedbacks between productivity, profitability and technology adoption.

agroecological approaches that are sensitive to supplies of local labour, organic inputs, improved knowledge and farm management (Pinstrup-Andersen, 2001).

Despite incomplete information about the best animal agriculture responses to global food demand drivers, and ways to motivate such implementations, short-term actions as well as research investments are needed. It is unrealistic that rural households can (or should) fully bear the burden of environmental costs because win-win opportunities are probably limited. Thus, the main challenge involves connecting existing knowledge about how agro-ecosystems function to the effectiveness of candidate interventions in specific situations. There are essentially five types of interventions that can be undertaken. These include financial incentives provided to farm households to promote desirable ecosystem outcomes (or reductions in subsidies for harmful activities); research to develop profitable technologies with minimal undesirable environmental impacts (or that are restorative); prohibition or restrictions on harmful activities with effective enforcement, and education for stakeholder groups about the ecological and environmental consequences (negative, positive) of their actions. In certain cases, innovative institutional arrangements (e.g., modification of land use rights, access to credit or banking that reduces dependence on livestock assets) will also be necessary or desirable. It is likely that some combination of these will be most effective. Land tenure, taxation and economic policies are other mechanisms for mitigation and restoration; and better communication is also needed so that science is better utilized in policy making (Fernandez, 2002).

To examine more fully the effectiveness of potential interventions, it will be helpful to integrate existing disciplinary knowledge into conceptual or quantitative simulation models. Modelling of agricultural systems has become an important way to assess the potential for policy interventions to prevent or mitigate environmental consequences (Thornton and Herrero, 2001; Heerink *et al.*, 2001). One general approach that can be usefully employed in this context is system dynamics (SD), a broadly applicable systems-based approach to problem solving (Sterman, 2000). Using SD, problems are expressed as a feedback (stock-flow) model that provides insights about how a problem behaviour developed over time and about which interventions or policies can provide a lasting solution to the problem. System dynamics is most appropriate for problems that are dynamically complex (evolve over time, are non-linear and governed by feedback), are long-term in nature, can be described by a relatively small set of "reference mode" behaviours observed in the real world, and can be described by flow processes (Vennix, 1996). The SD approach permits conceptual and quantitative modelling of dynamic problems, and has been previously applied to land degradation processes in sub-Saharan Africa (Brontkes, 2001). For successful application of these methods, it will be important to provide incentives to researchers from various disciplines, policy makers, development organizations, and farm households to work together to gain a broad perspective on the nature of both the problems and potential solutions. Because values, that is, the ultimate objectives to be achieved in response to the changes wrought by increased livestock demand, are likely to differ among the different disciplines (e.g., animal scientists, ecologists, economists), discussions of these values will be an important component of successful multidisciplinary research efforts.

Acknowledgements

We thank the following people for reading an earlier draft of this article and for their helpful suggestions to improve it: Duane Acker, G. Eric Bradford, Rod Heitschmidt and David Pimentel.

References

Alexandratos, N. (ed.) (1995). *World Agriculture towards 2010: An FAO study*. Chichester, UK: Food and Agriculture Organization of the United Nations and John Wiley and Sons Ltd.

Angelsen, A. and Kaimowitz, D. (2001). When does technological change in agriculture promote deforestation? In: *Tradeoffs or synergies? Agricultural intensification, economic development and the environment*. Edited by D.R. Lee and C.B. Barrett. CAB International Wallingford, UK, pp. 89-114.

Brontkes, T.S. (2001). Agricultural prices and land degradation in Koutiala, Mali: A regional simulation model based on farmer's decision rules. *Economic policy and sustainable land use: Recent advances in quantitative analyses for developing countries*. Edited by N. Heerink, H. van Keulen and M. Kuiper. Physica-Verlag, Heidelberg, pp.151-168.

Carpentier, C., Vosti, S.A. and Witcover, J. (2000). Intensified production systems on western Brazilian Amazon settlement farms: could they save the forest? *Agriculture Ecosystems and Environment* 82:73-88.

Cochrane, M. (2001). In the line of fire. Understanding the impacts of tropical forest fires. *Environment* 43:28-38.

Cranfield, J.A.L., Hertel, T.W., Eales, J.S. and Preckel, P.V. (1998). Changes in the structure of global food demand. *American Journal of Agricultural Economics* 80:1042-1050.

Delgado, C., Rosegrant, M., Steinfeld, H., Ehui, S., and Courbois, C. (1999). *Livestock to 2020: The next food revolution*. Food, Agriculture and Environment Discussion Paper 28, International Food Policy Research Institute (IFPRI), Washington, DC, USA.

Dobson, A.P., Bradshaw, A.D. and Baker A.J.M. (1997). Hopes for the future: Restoration ecology and conservation biology. *Science* 277:515-522.

Fernandez, R.J. (2002). Do humans create deserts? *Trends in Ecology and Evolution* 17:6-7.

Heerink, N., Kuyvenhoven, A. and van Wijk, M.S. (2001). Economic policy reforms and sustainable land use in developing countries: issues and approaches. *Economic policy and sustainable land use: Recent advances in quantitative analyses for developing countries*. Edited by N. Heerink, H.van Keulen and M. Kuiper. Physica-Verlag Heidelberg, pp. 1-20.

Kassa, H., Blake, R.W. and Nicholson, C.F. (2002). The crop-livestock subsystem and livelihood dynamics in the Harar Highlands of Ethiopia. In: *Responding to the increasing global demand for animal products - Programme and Summaries*. International conference organised by the British Society of Animal Science, American Society of Animal Science and Asociacion Mexicana de Produccion Animal, held at the Universidad Autonoma de Yucatan, Merida, Mexico, 12-15 November, 2002. British Society of Animal Production, Penicuik, UK, pp. 74-75.

Kauffman, J.B., Cummings, D.L. and Ward, D.E. (1998). Fire in the Brazilian Amazon. 2. Biomass, nutrient pools and losses in cattle pastures. *Oecologia* 113:415-427.

Keller, L.F. and Waller, D.M. (2002). Inbreeding effects in wild populations. *Trends in Ecology an Evolution* 17(5):230-241.

Laurance, W.F. (2001). Future shock: forecasting a grim fate for the Earth. *Trends in Ecology and Evolution* 16:531-533.

Laurance, W.F., Cochrane, M.A., Bergen, S., Fernside, P.M., Delamonica, P., Barber, C., D'Angelo, S. and Fernandes, T. (2001). The future of the Brazilian Amazon. *Science's Compass Policy Forum*, 19 January, 291:438-439.

Laurance, W. F., Powell, G. and Hansen, L. 2002. A precarious future for Amazonia. *Trends in Ecology and Evolution* 17:251-252.

Matson, P.A., Parton, W.J., Power, A.G. and Swift, M.J. (1997). Agricultural intensification and ecosystem properties. *Science* 277: 504-509.

Meadows, D.M. (2002). Reflections on the dynamics of growth. Paper presented at the Annual Meeting of the System Dynamics Institute, June 4, 2002, Montebello, Quebec.

McLaughlin, J.F., Hellmann, J.J., Boggs, C.L. and Ehrlich, P.R. (2002). Climate change hastens population extinctions. *Proceedings National Academy of Science* 99:6070-6074.

Nicholson, C.F., Blake, R.W., Reid, R.S. and Schelhas, J. (2001). Environmental impacts of livestock in the developing world. *Environment* 43:7-17.

Nicholson, C.F., Blake, R.W. and Lee, D.R. (1995). Livestock, deforestation, and policy making: Intensification of cattle production systems in Central America revisited. *Journal of Dairy Science* 78:719-734.

Oba, G., Stenseth, N.C. and Lusigi, W.J. (2000). New perspectives on sustainable grazing management in arid zones of sub-Saharan Africa. *BioScience* 50:35-51.

Perkins, J.S. (1996). Botswana: fencing out the equity issue. Cattleposts and cattle ranching in the Kalahari Desert. *Journal of Arid Environments* 33:503-517.

Pimentel, D., Lach, L., Zuniga, R. and Morrison, R. (2000). Environmental and economic costs of nonindigenous species in the United States. *BioScience* 50:53-62.

Pinstrup-Andersen, P. (2001). Feeding the world in the new millennium: Issues for the new US administration. *Environment* 43:22-31.

Rueda, B., Blake, R.W., Fernandes, E., Nicholson, C.F. and Valentim, J.F. (2004). Soil, plant and cattle nutrient pools on pastures of the western Amazon of Brazil. *Agriculture, Ecosystems and Environment* (submitted).

Rueda, B., Blake, R.W., Nicholson, C.F., Fox, D.G., Tedeschi, L., Pell, A.N., Fernandes, E., Lee, D.R., Valentim, J.F. and da Costa, J. (2003). Production and economic potentials of cattle in pasture-based systems of the western Amazon region of Brazil. *Journal of Animal Science* 81:2923-2937.

Sala, O.E., Chapin III, F.S., Armesto, J.J., Berlow, E., Bloomfield, J., Dirzo, R., Huber-Sanwald, E., Huenneke, L.F., Jackson, R.B., Kinzig, A., Leemans, R., Lodge, D.M., Mooney, H.A., Oesterheld, M., Poff, N.L., Sykes, M.T., Walker, B.H., Walker, M. and Wall, D.H. (2000). Global biodiversity scenarios for the year 2100. *Science* 287:1770-1774.

Sekercioglu, C.H., Ehrlich, P.R., Daily, G.C., Aygen, D., Goerhring, D. and Sandi, R.F. (2002). Disappearance of insectivorous birds from tropical forest fragments. *Proceedings National Academy of Science* 99:263-267.

Sterman, J.D. (2000). *Business dynamics: Systems thinking and modeling for a complex world*. McGraw-Hill, USA.

Tedeschi, L.O., Fox, D.G., Pell, A.N., Lanna, D.P.D. and Boin, C. (2002). Development and evaluation of a tropical feed library for the Cornell Net Carbohydrate and Protein Model. *Scientia Agricola* 59(1):1-18.

Thornton, P.K., and Hererro, M. (2001). Integrated crop-livestock simulation models for scenario analysis and impact assessment. *Agricultural Systems* 70:581-620.

Tilman, D. (1999). Global environmental impacts of agricultural expansion: The need for sustainable and efficient practices. *Proceedings National Academy of Science* 96:5995-6000.

Tilman, D., Fargione, J., Wolff, B., D'Antonio, C., Dobson, A., Howarth, R., Schindler, D., Schlesinger, W.H., Simberloff, D. and

Swackhamer, D. (2001a). Forecasting agriculturally driven global environmental change. *Science* 292:281-284.

Tilman, D., Reich, P.B., Knops, J., Wedin, D., Mielke, T. and Lehman, C. (2001b). Diversity and productivity in a long-term grassland experiment. *Science* 294:843-845.

Vennix, J.A.M. (1996). *Group model building: Facilitating team learning using system dynamics*. John Wiley and Sons, Chichester, UK.

Vitousek, P.M., Mooney, H.A., Lubchenco, J. and Melillo, J.M. (1997). Human domination of Earth's ecosystems. *Science* 277:494-499.

Vosti, S.A. (1995). Sustainability, growth, and poverty alleviation: Research and policy issues for tropical moist forests. *Agriculture and environment: Bridging food production and environmental protection in developing countries*. Edited by A.S.R. Juo and R. D. Freed. pp. 207-227. Special Publication 60, Soil Science Society of America.

Vosti, S.A. and Valentim, J. (1998). Foreward. In: *Cattle, deforestation and development in the Amazon: An economic, agronomic and environmental perspective*. (M.D. Faminow, Ed.) CAB International, Wallingford, UK, pp. vi-viii.

Wackernagel, M., Schulz, N. B., Deumling, D., Linares, A. C., Jenkins, M., Kapos, V., Monfreda, C., Loh, J., Myers, N., Norgaard, R. and Randers, J. (2002). Tracking the ecological overshoot of the human economy. *Proceedings National Academy of Science* 99:9266-9271.

Western, D. (2001). Human-modified ecosystems and future evolution. *Proceedings National Academy of Science* 98:5458-5465.

White, G. M., Boshier, D. H. and Powell, W. (2002). Increased pollen flow counteracts fragmentation in a tropical dry forest: An example from *Swietenia humilis* Zuccarini. *Proceedings National Academy of Science* 99:2038-2042.

10

Demand-driven crop-ruminant intensification: trans-regional analysis to understand patterns of change using village level data from three continents

I. Baltenweck, S. Staal and M.N.M. Ibrahim
International Livestock Research Institute (ILRI), PO Box 30709, Nairobi 00100, Kenya

Introduction

The growing demand for livestock products in the developing world is expected to drive livestock production in these countries. Delgado *et al.* (1999) predict that by 2020, developing countries will produce on average 38% more meat and 54% more milk *per capita* than in the early 1990s. The livestock revolution could then translate into opportunities for smallholder livestock producers. In the case of ruminants, much of this increase may occur in mixed farming systems, rather than through industrial production. In 1991-93, mixed systems produced 92% of the world's milk, and 54% of the world's meat (de Haan *et al.*, 1997).

The attention given to these demand-led trends, however, has also underlined the fact that the precise nature of the resulting expected intensification in mixed systems is unclear. In general, the wide variety of ruminant production systems observed globally, differing in both scale and level of intensity, points to the importance of local factors in determining a farmer's production choices. Those factors may include not only local agro-climatic conditions and animal disease challenge, but also infrastructure, access to markets, capital and services, nature of local demand, the opportunity costs of labour, but critically on the potential for interaction with crop production and other uses of land. Given the complexity of outcomes and the certainty of coming demand-driven intensification, researchers and planners are currently poorly equipped to predict where and how changes will occur in crop-ruminant production systems. While the basic driving factors can be identified, the understanding of their effects is not adequate to make predictions elsewhere and for the future. In an attempt to address some of the issues, this paper presents the analysis of crop- ruminant intensification and interaction at a broad dimension, using farmer-group interviews from sites in 15 countries.

Theoretical background

The general model of crop-livestock interactions and intensification originates in the work of Boserup (1965) and McIntire *et al.* (1992). Intensification is described as an endogenous process in response to increased population pressure. As the ratio of land to population decreases, farmers are induced to adopt technologies that raise returns to land at the expense of a higher input of labour. The direct causal factor is relative factor price changes, in accordance with the theory of induced innovation (Hayami and Ruttan, 1985).

Besides population growth, other factors have been identified in the literature as determinants of the structure of crop-livestock systems. Environmental characteristics play a significant role in determining the evolution of crop-livestock systems (McIntire *et al.*, 1992; Powell and Williams, 1993). In humid areas, with a high disease challenge for large ruminants, levels of interaction are likely to be low since livestock are fewer. Other driving factors mentioned in a number of studies (e.g. Powell and Williams, 1993) are economic opportunities, cultural preferences, climatic events (e.g. drought that leads to livestock losses), lack of capital to purchase animals and labour bottlenecks at some periods of the year (e.g. harvests) that may prevent farmers from adopting technologies like the use of draught power.

Crop-livestock interactions are sometimes seen as an evolutionary process. As first described by McIntire *et al.* (1992) the effect of population density on crop-livestock interactions can be described as an inverted 'U' relationship. As population density increases, the level of interaction increases and reaches a maximal level at an intermediate level of population density, after which specialization and less interaction occurs at higher density. In fact, at low levels of population density, crop and livestock production systems are extensive and the sole interactions are through markets and contracts (e.g. manure contracts whereby feed is exchanged for manure). With population growth, systems intensify due to changing relative factor prices. Both the net demand for agricultural products and the opportunity costs of land increase, bringing about the need for on-farm crop-livestock interactions, mainly through more efficient exploitation of nutrient resources, crop residues, manure and the use of draught power. Because of transaction costs and other market constraints, on farm interactions are preferred to exchanges through contracts. At higher levels of intensification, markets develop and crop-livestock interactions become less attractive. There is a return to specialization and dependence on purchased inputs.

However, the above characterization of intensification and interaction does not adequately reflect several confounding factors, namely agro-climate and income levels. In low potential climatic zones, the final stages of specialization may not occur. Further, in the Boserup (1965) framework higher population densities are associated with relatively lower labour opportunity costs. However, the McIntire *et al.* (1992) framework of specialization at the final stages implies both increasing human population density and rising labour opportunity costs that drive the need to substitute capital for labour. One aim of this analysis is thus to differentiate the effects of population density changes with those of changes in relative labour opportunity costs.

Presentation of the sites

A total of 48 sites in 15 countries of Sub-Saharan Africa, Asia and Latin America were surveyed from November 2000 to October 2001. The countries surveyed include:

Asia: Thailand, Nepal, Bangladesh, Sri Lanka and India (Gujarat and Pondicherry)
Africa: Kenya, Tanzania, Ethiopia, Madagascar, Nigeria and Niger
Latin America: Bolivia, Columbia, Costa Rica and Peru

The data collection was carried out in half- day or one-day visits to a specific area. Farmer- group interviews were used to collect the needed village-level data, using a checklist. Information gathered included:

- Basic information (annual rainfall, cropping seasons, human and ruminant population density, distance to major urban centres, and assessment of the road infrastructures)
- General information (food sources and food habits, sources of income, off-farm opportunities, education level, access to facilities, and availability of agricultural equipment)
- Systems of production (land holding, crops and cropping seasons, livestock population including species/breed, land rental, market, and animal traction)
- Indicators of intensification (cattle/buffalo keeping strategies, mechanization, use of hired labour, indicators of milking performance, forage or crop residue markets, use of hybrid/high yield varieties or seeds, pressure on land, inorganic fertilizer use, animal manure use and market).

Sites were selected to reflect a wide range of climatic characteristics (Figure 10.1) and level of agricultural intensification. A total of 18 sites were surveyed in the semi-arid zone, 8 in the sub-humid zone, 6 in the humid area and 16 in the highlands. Semi-arid, sub-humid and humid areas are defined as areas with 50 to 180, 180 to 270, and more than 270 days of growing time, respectively. Highlands are defined as sites over 1,000 m above sea level. This definition

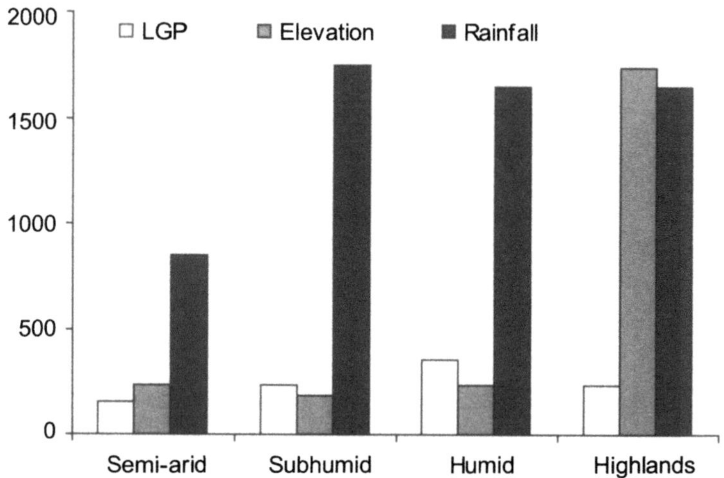

Figure 10.1. Climate characteristics of the surveyed sites, by climatic zone: average length of growing period (LGP, days), elevation and rainfall.

follows Jahnke (quoted by McIntire *et al.*, 1992). The analysis combines data from both the village data collection and GIS data layers, following the method developed by Staal *et al.* (2002). Data on length of growing period and on human population density are extracted from the global ILRI GIS database (Fisher *et al.*, (2000) cited in Thornton *et al.*, (2002), for the length of growing period; Deichman (1996a), Deichman (1996b) and Hyman *et al.*, (2000) cited in Thornton *et al.*, (2002), for human population densities).

Cattle, irrespective of breed, are the main type of livestock kept in the villages studied (Table 10.1). In all the areas, except the humid zones, local breeds of cattle are the most common livestock. Crossbred cattle are defined as crosses between a local breed and a European dairy breed. Goats and sheep are very common in the semi-arid areas. Other studies (Powell and Williams, 1993) showed the importance of small stock in these areas, due to their lower feed requirements and lower costs. As expected, high-grade cattle are more common in the highlands, due to the lower incidence of disease. Data on buffalo are calculated only for the sites where buffalo are present.

Table 10.1. Percentage of farmers with livestock, by climatic zone.

	Cattle		*Buffalo*					
Climatic zone	*Local breed*	*Cross-bred*	*High grade*	*Local breed*	*Dairy*	*Goat*	*Sheep*	*Donkey*
Semi-arid, no irrigation	60	9	8	8	37	64	41	10
Semi-arid, irrigation	46	9	25	28	26	32	16	14
Sub-humid	47	31	26	16	4	33	0	6
Humid	32	57	28	60	45	5	11	30
Highlands	45	13	40	37	27	17	28	20

Data analysis

Driving forces

Building on the literature, four driving forces of intensification and crop-livestock interactions are identified:

1. The general level of education in the villages (percentage of farmers with at least primary education) captures the average labour productivity in the area. This follows a common approach taken in the literature; Feder *et al.* (1985) show that the education level is usually positively associated with the adoption of agricultural technologies in developing countries since it captures farmers' abilities to manage new, and possibly, complex technologies. Since extension services can be seen as a substitute for education, the availability of extension services in the village is also used to capture labour productivity.
2. Another driving force is the existence of labour opportunities in the neighbourhood: farmers with such opportunities may be less induced to invest in agricultural activities, resulting in more extensive systems of production.
3. The third determinant of intensification is the proximity to urban centres where demand for agricultural products is concentrated: farmers located near urban centres experience lower transaction costs resulting in lower inputs prices and higher output prices. Therefore, the closer the village to an urban centre, the higher the incentives to intensify livestock production.
4. Costs of labour and land constitute the fourth set of driving forces: farmers with lower labour costs are less likely to intensify due to higher production costs.

Three indicators are used in the analysis: the monthly casual labourer wage rate in US dollars; the cost of renting land (US$/ha per year); and the ratio of the wage rate over land rental rate. Costs of labour and land are expressed in two ways: in current US$ (local rates converted in US$ using market exchange rate) and in Purchasing Power Parity (PPP) US$ (local rates converted into US$, using rates of conversion that equalize the purchasing power of different currencies by eliminating the differences in price levels between countries, OECD Website). Following Boserup's (1965) work, human population density is another driving force introduced into the analysis. Lastly, climatic characteristics are captured using the length of growing period (number of days).

Ruminant intensification

The first indicator of ruminant intensification is the feeding system that is the most common in the village. Four types are distinguished:

grazing (unimproved and improved) pasture; mainly grazing (including tethering) with some stall-feeding; mainly stall-feeding with some grazing; and stall-feeding only. Table 10.2 presents the distribution of the surveyed sites by climatic zone and main type of feeding system. It is expected that production systems will be extensive in the semi-arid areas although data only partly confirm this hypothesis, since the type of feeding system in semi-arid areas is grazing for 7 sites out of 18; however, for half of the sites animals are either mainly or entirely stall-fed. Other driving forces thus seem to be at play.

Table 10.2. Number of sites by climatic zone and feeding system (numbers in parentheses indicate the percentage of sites within climatic zone).

	Semi-arid	*Sub-humid*	*Humid*	*Highlands*	*Total*
Grazing	7 (39)	2 (25)	1 (16)	4 (25)	14 (29)
Mainly grazing	2 (11)	3 (38)	4 (67)	3 (19)	12 (25)
Mainly stall-feeding	4 (22)	1 (12)	0 (0)	3 (19)	8 (17)
Stall-feeding	5 (28)	2 (25)	1 (17)	6 (37)	14 (29)
Total	18	8	6	16	48

The extent of land that farmers allocate to planted forage (defined as a crop exclusively for livestock feeding) is another indicator of livestock intensification. Intensification levels are considered low in sites where farmers plant forage on less than 5% (more than 5% indicates high intensification) of the cropped land. The threshold of 5% was chosen for two reasons: firstly, planting forage on less than 5% of the land indicates that farmers are at the experimental stage of planting forage, or the forage, is only marginally used as feed. Secondly, using 5% as the threshold allowed getting a sufficient number of sites in each category. The other indicator of ruminant intensification is the percentage of farmers who purchase concentrates, i.e. a feed aimed at increasing animals' performance. Sites where less than 50% of the farmers purchase concentrates have low levels of intensification (more than 50% indicates high intensification). Table 10.3 presents the results of the statistical analysis (mean comparison tests between groups based on Student's distribution statistics), differentiating sites based on these different indicators.

For the three indicators, the general level of education in the village is significantly higher in the most intensified sites. In fact, the level of education captures farmers' ability to adopt new technologies associated with the intensification process. However, there is no apparent relationship between access to livestock extension services and intensification level. As expected, intensification is significantly lower in sites with more off-farm opportunities, although only when

Table 10.3. Determinants of ruminant intensification on feeding system (grazing vs. stall-feeding), percentage of land in forage (<5 vs. > 5%) and percentage of farmers purchasing concentrates (<50 vs. >50%) (mean comparison tests between groups).

Variable	Feeding system			% land in planted forage			% farmers purchasing concentrates		
	Grazing	Stall-feeding	Sig.	<5%	>5%	Sig.	<50%	>50%	Sig.
Number of sites	26	22		29	19		20	28	
% farmers with primary education	59	77	**	62	75	*	54	77	***
% farmers with livestock services	62	69		70	57		57	71	
% farmers with off-farm income	15	19		20	12	*	16	17	
Distance to nearest big city (km)	72	27	**	46	61		79	32	**
Casual wage rate (current $)	66	33	**	44	60		55	47	
Casual wage rate (PPP $)	178	140		154	170		145	170	
Land rent (current $)	131	191		147	173		54	225	***
Land rent (PPP $)	523	882	*	671	724		218	1011	***
Wage rate/ rent	1.56	0.34	***	1.33	0.45	**	2.02	0.35	***
Human population density (persons/km^2)	267	344		322	273		193	380	***
Length of growing period (days)	222	203		198	237	*	191	229	*

*, **, *** The difference across groups is statistically significant at $P < 0.100$, 0.050 and 0.010, respectively

of mixed crop-ruminant systems in developing countries. The results described here show that those changes are occurring in consistent patterns. Although agro-climatic characteristics are important to understand smallholder crop-livestock systems, socio-economic factors do play a role in explaining crop-livestock intensification and interaction levels. Costs of the primary factors of production, land and labour, determine partly the structure of crop-livestock smallholder production systems. Also, human population density, education level and access to market are important factors. Key findings are those that relate choice of livestock practices to relative labour and land costs, and to market access. For example, use of planted forage is closely linked to low opportunity costs of labour relative to costs of land. However, this understanding is rarely applied to forage promotion efforts, which typically recommend forage cultivation and species selection based exclusively on agro-climatic zone, and do not recognize the strong negative effect of factor costs. Similar results are found for stall-feeding systems that also depend heavily on low labour costs. However, village-level results hide heterogeneity at the farm level. Next steps of the trans-regional project aim at analyzing smallholder crop-livestock systems at the landscape and farm-levels, to decompose broad patterns into underlying components.

Acknowledgements

This publication is an output from the Trans- Regional Analysis of Crop-Livestock Systems, a project led by the International Livestock Research Institute (ILRI) and funded by the System-wide Livestock Programme (SLP). The authors thank particularly Victor Manyong, Federico Holman, B.R. Patil and P.K. Thornton for their comments.

References

Boserup, E. (1965). *The condition of agricultural growth: the economics of agrarian change under population pressure*. Allen and Unwin, London, UK, 108 pp.

Delgado, C., Rosegrant, M., Steinfeld, H., Ehui, S., and Courbois, C. (1999). *Livestock to 2020: The next food revolution*. Food, Agriculture and Environment Discussion Paper 28, International Food Policy Research Institute (IFPRI), Washington, DC, USA, 72 pp.

Feder, G., Just, R.E. and Zilberman, D. (1985). Adoption of agricultural innovations in developing countries: a survey. *Economic Development and Cultural Change* 33: 255-298.

Haan, C. de, Steinfeld, H. and Blackburn, H. (1997). *Livestock and the environment: Finding a balance*. Wrenmedia, Suffolk, UK.

Hayami, Y., and Ruttan, V.W. (1985). *Agricultural development: an international perspective*. Revised edition, The John Hopkins University Press, London, UK, 528 pp.

McIntire, J., Bourzat, D. and Pingali, P. (1992). *Crop-livestock interaction in Sub-Saharan Africa*. The World Bank Washington, DC, 244 pp.

OECD Website. http://www.oecd.org/std/ppp/pppfaq.htm

Powell, J.M. and Williams, T.O. (1995). An overview of mixed farming systems in sub- Saharan Africa. In: *Livestock and sustainable nutrient cycling in mixed farming systems of sub-Saharan Africa. Volume 11: Technical papers*. Proceedings of an International Conference held in Addis Ababa, Ethiopia, 22-26 November 1993, Edited by Powell, J.M., Fernandez-Rivera, S., Williams, T.O. and Renard, C. International Livestock Centre for Africa (ILCA), Addis Ababa. pp. 21-36.

Staal, S.J., Baltenweck, I., Waithaka, M.M., de Wolff, T. and Njoroge, L. (2002) Location and uptake: integrated household and GIS analysis of technology adoption and land use, with application to smallholder dairy farms in Kenya. *Agricultural Economics* 27: 295-315

Thornton, P. K., Kruska, R.L., Henninger, N., Kristjanson, P.M., Reid, R.S., Atieno, F., Odero, A.N. and Ndegwa, T. (2002). *Mapping poverty and livestock in the developing world*. International Livestock Research Institute (ILRI), Nairobi, Kenya, 124 pp.

11

Effect of intensification on feed management of dairy cows in the Central Highlands of Kenya

D. Romney[1], C. Utiger[1,4], R. Kaitho[1], P. Thorne[3], A. Wokabi[2], L. Njoroge[1], L, Chege[2], J, Kirui[2], D. Kamotho[2] and S. Staal[1]
[1]*International Livestock Research Institute (ILRI), PO Box 30709, Nairobi, Kenya*
[2]*Ministry of Agriculture and Rural Development, PO Box 30082, Nairobi, Kenya*
[3]*Stirling Thorne Associates, PO Box 23, Llangefni, Ynys Mon LL74 8ZE, UK*
[4]*Institute of Animal Sciences, ETH, CH-8092 Zurich, Switzerland*

Introduction

In sub-Saharan Africa mixed crop-livestock systems predominate in the semi-arid, sub-humid and cool highland zones. In these areas, systems intensify and crops and livestock become increasingly integrated as the human population increases and land becomes a more important constraint than labour (Boserup, 1965; McIntire *et al.*, 1992). As intensification progresses, use of crop residues moves from open access to crop fields, following harvest, to labour intensive management of cereals as dual-purpose crops. In some cases, farmers harvest thinnings and green stover during the growing period as well as harvesting and storing the residues to feed to confined animals (Lukuyu, 2000). Manure becomes a vital source of crop nutrients as scarcity of land precludes the use of fallow. Together with the stimulus of increasing demand for livestock products (Delgado *et al.*, 1999), the trend is for systems to intensify, with output per unit of land increasing (Staal *et al.*, 2000a). In the Central Highlands of Kenya, smallholder dairy systems provide livelihoods for more than 50 per cent of agricultural households (Staal *et al.*, 2002a). Farming systems are continually evolving in response to changes in available resources and access to market. In order to develop appropriate technologies and target extension advice it is important to understand how intensification influences the management strategies that farmers adopt. This paper describes a study of feeding strategies in two areas of Kenya where population densities varied and farming systems were considered to represent different levels of intensification. A longitudinal study was carried out where a total of 42 farms were visited regularly during a period of 14 months in order to capture seasonal effects in feed management, production and markets.

Materials and methods

Study site and farm selection

Farms were selected from Central and Rift Valley Provinces (Figure 11.1) to represent areas of high (HI) and low (LI) levels of intensification. Selection of the 21 farmers in each area was stratified to represent important systems observed during earlier characterization surveys in the region (Staal *et al.*, 2002). In Central province (HI), in Kiambu District to the North West of Nairobi, farms were selected where coffee (n=11) or horticulture (kale, cabbage and green maize (n=10) were the main cash crops. In Rift Valley Province (LI) farmers were selected in Nakuru (n=11) and Nyandarua (n=10) Districts, where population density was lower compared to Kiambu and farm sizes generally larger. Main cash crops were horticultural (cabbages and kales) in Nakuru, with pyrethrum important in Nyandarua as well as cabbages and potatoes. Market access was poorest in Nyandarua where there were problems of marketing milk, particularly in the rainy season, reflected by lower average milk prices (17.6, 15.2 and 14.3 KSh/l in Kiambu, Nakuru and Nyandarua, respectively, at the time of the surveys).

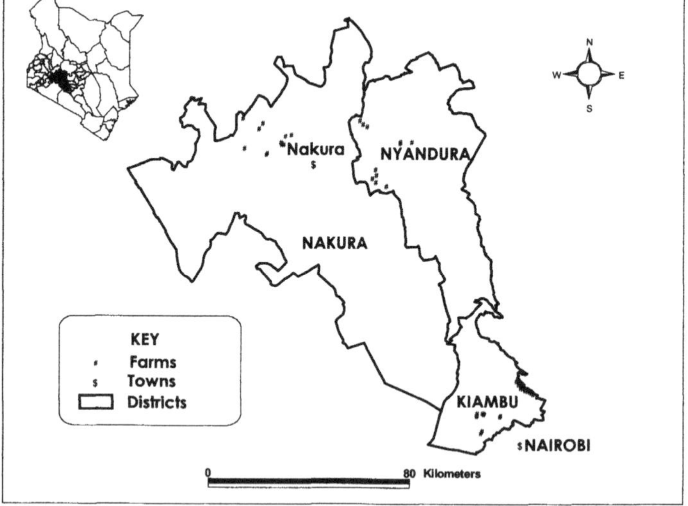

Figure 11.1. Location of selected farms in Nakuru, Nyandarua and Kiambu. Nakuru and Nairobi towns are also marked.

Data collection

In order to characterize the different systems, household parameters were collected including land size and number of cattle owned. All farms were geo-referenced and GIS layers used to estimate population density in a 5km radius of each farm, and market access was described as the total time to drive to the nearest market centre, taking account of different road types (Staal *et al.*, 2002b). All farmers were visited twice weekly over a period of 14 months to collect data on feeding strategies and milk yield. Data collection took place from November 1997 to December 1998 in Central Province and from January 1999 to February 2000 in Rift Valley Province. Once a fortnight enumerators spent a full day on each farm recording quantity, type and source (on or off-farm) of feed offered to cattle during the day, whether or not the feed coming from off-farm was purchased and milk yield. Annual supply of offered dry matter (DM) from concentrate, crop residues, planted fodder, grazed pasture and other cut and carry fodders was estimated relative to the TLUs Tropical Livestock Units (TLU = 250 kg) present on the farm across the year.

Data analysis

Characteristics related to individual households with no seasonal effects, including local population density, market access, land size and cattle number, were analyzed using a mixed model (PROC MIX in SAS) where Intensification Group (IG) and System within group were considered as fixed effects with farm being the random effect. Milk production parameters were also analyzed in this way, with production per hectare representing an indicator of level of intensification. Since unusual rainfall patterns occurred during the monitoring period, seasonal effect on feeding patterns was examined by estimating amounts of feed offered during months where rainfall was above (wet) or below (dry) the overall average, according to rainfall measurements from meteorological stations in Kiambu, Nakuru and Nyandarua. In the HI areas and in Nakuru and Nyandaurua Districts (LI), 7, 2 and 5 months out of 12 were above average respectively. These data were also analyzed using a mixed model but with intensification group (IG), season (S), system within IG and IG by S interaction fitted as fixed effects, with the random effect being farm within systems nested within groups. The first 2 months data were not used in the analysis since they were taken as lead-in periods when enumerators were learning the data collection procedures.

Results and discussion

Household characteristics and intensification indicators

Table 11.1 shows that population density in HI areas was more than six times higher than that in LI areas, reflected by smaller farm sizes ($P<0.05$). Numbers of cattle owned were lower in HI areas compared to LI areas ($P=0.060$) as expected, however, because of differences in land sizes, stocking density in Nakuru district was similar to those in the intensive sites. Market access was greater in HI ($P<0.001$)

with no difference between the sub-groups. Within LI areas market access was significantly poorer ($P<0.001$) in Nyandarua. The indicator of time taken to the nearest market centre generally reflects the difficulty with which farmers sell milk, since it describes the remoteness of the farm and the difficulty with which milk can be collected or taken to market. Market access was expected to have a significant effect on management choices at the farm level. For example in Nyandarua, although some processors collect milk this tends to be in times of scarcity when they cannot source milk from more accessible farms or local markets. A number of farmers report throwing away milk during the season when milk production elsewhere in the country is at a peak because they were unable to sell it or consume it at home. Poor market access was reflected by a lower proportion of milk produced being sold in Nyandarua (57% compared to 70 and 71% in Kiambu and Nakuru respectively) (SDP unpublished reports). Milk production per farm or TLU is similar in all systems ($P<0.05$). However, the intensification indicator, output per unit of land (l/ha), increases as farm size decreases and farmers intensify dairy production. These changes reflect use of on and off-farm feed resources, as discussed in the following sections.

Effect of intensification

Analysis of differences between systems within group indicated no significant differences, reflecting the non-significant difference within group for intensification indicators such as milk production/ha (Table 11.1). Therefore, only main effects of intensification and season are reported (Table 11.2). Total amount of DM offered per animal was slightly higher in the more intensive area (HI), although the difference did not quite reach significance ($P=0.67$). Significant differences were observed in the type and source of feed dependant on the level of intensification. In the LI system, where farming was more extensive, the main source of feed was from grazed pastures on farmers' own land and from communal grazing sources off-farm. In the more intensive HI systems, grazing was almost non-existent with almost all animals being stall-fed, and receiving much higher amounts of the planted fodder, Napier grass (*Penniseteum purpureum*) ($P<0.001$), and concentrate ($P<0.001$) than in the LI system. In intensive systems some concentrate was produced on-farm in the form of poultry litter, while in extensive systems the little concentrate offered was all purchased. Crop residues were offered in similar amounts ($P>0.05$) and contributed a similar proportion of total feed (26 and 27% for HI and LI respectively). However, in the LI system, with an average of 1.8 ha under maize, crop residues were mainly from maize with some vegetable and vegetable waste, mainly potatoes and cabbage. In the HI system where the area under maize was much smaller (0.19 ha) the proportion of maize residues offered tended to be lower

Table 11.1. Household parameters in areas of high (HI) or low (LI) intensification. Least square means are represented and means with different superscripts differ significantly (*P*<0.05=*; <0.01 = **; <0.001 = ***; >0.05 = NS).

Intensification group (IG)	*HI*		*LI*		*sed*		*Sig*	
Agro-ecological zone defined as accompanying cash crop	*Hortic.*	*Coffee*	*Nakuru Hortic./ coffee*	*Nyandarua Coffee/ wheat/ sheep*	*IG*	*System within group*	*IG*	*System within group*
Population density (persons /km^2)	763^a	617^b	101	96	24.3	34.3	***	***
Market access (minutes to market centre)	14.4	13.7	21.0^a	45.4^b	1.94	2.75	***	***
Land size (ha)	0.8^a	1.7^{ab}	2.9^{ab}	4.8^b	1.18	1.66	*	NS
Cattle owned (TLU)	3.3	3.7	6.2	5.3	1.19	1.68	0.060	NS
Stocking rate (TLU/ha)	4.8^a	3.5^{ab}	4.5^a	1.7^b	0.90	1.23	NS	0.076
Milk Production (l/day/TLU)	2.3	2.2	1.8	2.4	0.39	0.55	NS	NS
Milk Production (l/farm/day)	7.2	8.9	13.1	11.3	3.47	4.90	NS	NS
Milk production (l/day/ha)	12.3^a	9.2^{ab}	7.1^{ab}	4.0^b	2.27	3.21	*	NS

TLU = Tropical Livestock Unit = 250 kg liveweight

Table 11.2. Total quantity of feed dry matter (DM) offered and the amount offered from off-farm sources and the proportion of the off-farm feed that was bought. Values are adjusted means for the wet and dry months in areas of high (HI) and low (LI) intensification. Effect of intensification group (IG), season (S) and the interaction between the two are presented.

	HI		*LI*		s.e.d			*Significance*		
	Wet	*Dry*	*Wet*	*Dry*	*IG*	*S*	*IGXS*+	*IG*	*S*	*IGXS*
Total feed, kg DM / TLU / day										
All feed	7.8	8.3	6.0	7.1	0.82	0.33	0.44-0.92	0.067	*	NS
Concentrate	0.90	1.58	0.23	0.23	0.22	0.12	0.16-0.27	***	**	**
Crop residues	1.6	2.5	1.1	2.4	0.39	0.30	0.40-0.55	NS	***	NS
Other cut/carry forage	0.6	0.7	1.1	1.0	0.28	0.10	0.13-0.30	NS	NS	NS
Grazing	0.0	0.1	2.8	2.5	0.25	0.07	0.09-0.27	***	*	*
Planted forage	4.7	3.5	0.8	1.0	0.46	0.18	0.23-0.50	***	**	***
On-farm feed, kg DM / TLU / day										
All feed	5.7	4.9	3.4	4.3	0.66	0.30	0.40-0.76	*	NS	**
Concentrate	0.08	0.08	0.00	0.00	0.03	0.01	0.02-0.04	*	NS	NS
Crop residues	1.3	1.5	0.7	1.6	0.28	0.24	0.31-0.41	NS	*	NS
Other cut/carry forage	0.4	0.4	0.3	0.4	0.13	0.06	0.09-0.15	NS	NS	NS
Grazing	0.0	0.1	1.6	1.4	0.32	0.06	0.08-0.33	***	0.067	*
Planted forage	3.8	2.9	0.7	1.0	0.38	0.18	0.24-0.44	***	*	***
Off-farm feed, kg DM / TLU / day										
All feed	2.2	3.4	2.6	2.7	0.69	0.24	0.31-0.75	NS	**	*
Concentrate	0.82	1.50	0.23	0.23	0.23	0.12	0.16-0.27	***	**	**
Crop residues	0.27	1.01	0.38	0.80	0.27	0.18	0.24-0.35	NS	***	NS

Table 11.2. Contd.

	HI		*LI*		*s.e.d*			*Significance*		
	Wet	*Dry*	*Wet*	*Dry*	*IG*	*S*	*IGXS*⁺	*IG*	*S*	*IGXS*
Other cut/carry forage	0.21	0.32	0.78	0.60	0.26	0.07	0.09-0.27	NS	NS	*
Grazing	0.00	0.02	1.16	1.07	0.24	0.05	0.06-0.25	***	NS	NS
Planted forage	0.85	0.58	0.07	0.04	0.22	0.09	0.12-0.25	**	NS	NS
Purchased feed										
All feed (kg DM / TLU / day)	2.0	3.1	0.7	1.1	0.59	0.23	0.31-0.66	**	***	NS
All feed (proportion of off-farm)	0.90	0.88	0.42	0.40	0.08	0.02	0.03-0.08	***	NS	NS
Concentrate	All purchased									
Crop residues	All purchased									
Other cut/carry forage (proportion of off-farm)	0.40	0.40	0.06	0.07	0.10	0.04	0.05-0.11	***	NS	NS
Grazing	None purchased									
Planted forage	All purchased									

⁺A range is given for s.e.d for the interaction. This is because differences in group size make it inappropriate to report mean s.e.d.
TLU = Tropical Livestock Unit (250kg)

(P=0.07), with one quarter of residues coming from banana, sweet potatoes and other vegetables, with proportions as high as 0.60 being observed

Little difference in the total amount of feed DM from off-farm sources was observed between the two study sites (P=0.07). However, in the extensive system this represented grazing, crop residues and other cut and carry forage (cut grass, hay, kitchen waste and vegetables). In intensive systems, although the amount of crop residue was similar, there was almost no grazing and relatively large amounts of planted forage and concentrates offered (0.71 and 1.16 kg DM/ TLU/day, respectively). Concentrates, crop residues, and planted forage from off-farm sources were always purchased, while pasture was always free. This resulted in a larger proportion of purchased feed in the intensive system (P<0.001) where almost all of the feed from off-farm sources was purchased.

Seasonal variation

In both intensification groups more feed was offered in the dry months (P<0.05), however the strategies used to achieve higher offer rates in the two areas were very different. In the HI system farmers compensated for a lower amount of feed being produced on farm, with decreases in total feed DM largely resulting from decreased amounts of planted forage (P<0.05), by increasing the amount of both concentrates and crop residues purchased off-farm (P<0.01). In contrast, in the LI system there was an increase in the amount of feed produced on-farm with increased quantities of crop residues offered (P<0.05). The total amount of feed in the LI system sourced off-farm did not change, but an increase in the amount of crop residues purchased and a small decrease in all other feeds was observed. The amount of concentrate offered in the LI system was low in both wet and dry months

Discussion

McIntire *et al.* (1992) proposed that the effect of population density on crop-livestock interactions could be described by an inverted U shaped relationship. Where low population densities exist, farmers specialise in crop or livestock production with cattle under extensive management. Under these conditions feed scarcity is overcome by cattle mobility and crop farmers maintain soil fertility by the use of long-term fallows. As land becomes scarcer crop and livestock enterprises become integrated on the same farm, with nutrient recycling improving efficiency of nutrient capture (McIntire *et al.,* 1992). In terms of feed resource use this relationship can be represented by a series of inverted U-shaped curves for different feed resources with crop residues increasing and then declining in importance as planted fodder and supplements replace grazing at high population densities (Figure 11.2, S. Fernandez-Rivera, personal communication). Increasing levels of planted forage and concentrate use are also considered as intensification indicators (see Chapter 10). Patterns observed in the present study can be considered to conform to this pattern, with crop residues contributing approximately 25% of feed DM offered in both systems, with natural grazing in the LI system (41% of feed DM) being replaced by planted forage in the HI system (51% of feed DM). Planted forages are usually rare as sources of feed even where there are well-adapted species available (McIntire *et al.*, 1992). These authors give three reasons for this: 1) the competition between land and labour; 2) the seasonality of planted fodders which are abundant at the same time as alternative, cheaper sources; and 3) the mobility of livestock, allowing them to move to harvest cheaper resources on common access land. The Kenyan Highlands is, however, cited as an exception where good market access results in good returns to investment in these materials. At the same time limited common lands and almost universal confinement of cattle limit the alternatives.

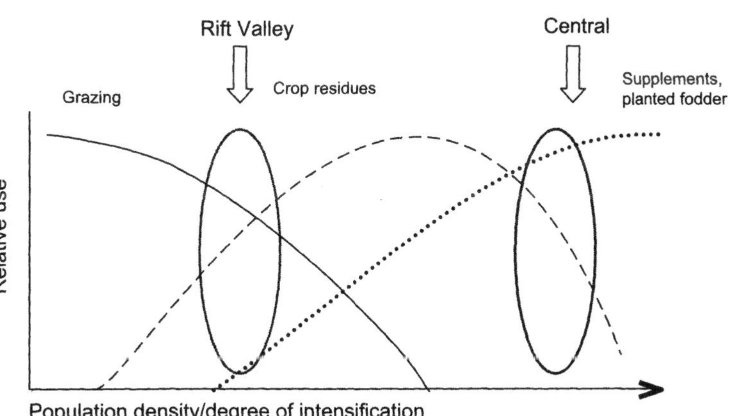

Figure 11.2. Effect of intensification on feeding strategies (*Personal communication: S Fernandez-Rivera, ILRI*).

Interactions between crop and livestock enterprises on different farms are considered mainly in extensive systems, for example, where crop farmers allow grazing of residues in order to obtain animal manure (Powell and Williams, 1995). However, in the present study 20% of total forage offered is purchased from other farmers in the HI system, compared to only 10% in the LI system. This suggests that in the study area, despite very high population densities, farmers preferred to maintain a mixed enterprise rather than specialize. Thorne *et al.* (2002) suggest there may be cultural resistance to replacing a staple such as maize with Napier grass, which is effectively a cash crop.

Conclusions

It was clear from the present study that farmers use a wide variety of feeds to cover livestock requirements. At very high population densities farmers do not necessarily specialise but maintain a mixed enterprise by purchasing forages from other farmers as well as offering relatively high levels of concentrate feeds. Recommendations to specialize further by increasing the area under planted forages in such situations may be resisted by farmers who are able to source feed off-farm.

Acknowledgements

This publication is an output from a research project (Livestock Production Programme) funded by the UK Department for International Development (DFID) for the benefit of developing countries. The views expressed are not necessarily those of DFID.

References

Boserup, E. (1965). *The condition of agricultural growth: the economics of agrarian change under population pressure*. Allen and Unwin, London, 108 pp.

Delgado, C., Rosegrant, M., Steinfeld, H., Ehui, S., and Courbois, C. (1999). *Livestock to 2020: The next food revolution*. Food, Agriculture and Environment Discussion Paper 28, International Food Policy Research Institute (IFPRI), Washington, DC, USA., 72 pp.

Jaetzold, R. and Schmidt, H. (1983). *Farm management handbook of Kenya*. Ministry of Agriculture, Kenya.

Lukuyu B.A. (2000). *The maize crop as a source of food and feed for livestock on smallholder dairy farms in the Kenyan Highlands*. MPhil Thesis, The University of Greenwich.

McIntire, J., Bourzat, D. and Pingali, P. (1992). *Crop-livestock interaction in sub-Saharan Africa*. Washington D C., USA, 244 pp.

Powell, J.M. and Williams, Y. (1995). An overview of mixed farming systems in sub-Saharan Africa. In: *Livestock and sustainable nutrient cycling in mixed farming systems of sub-Saharan Africa. Volume 11: Technical Papers*. Proceedings of an International Conference held in Addis Ababa, Ethiopia, 22-26 November 1993, Edited by Powell, J.M, Fernandez-Rivera, S., Williams, T.O. and Renard, C. ILCA (International Livestock Centre for Africa), Addis Ababa, pp. 21-36.

Staal, S. J., Ehui, S. and Tanner J. (2000). Livestock-environment interactions under intensifying production. In: *Tradeoffs or synergies? Agricultural intensification, economic development and the environment*. Edited by Lee, D.R. and Barrett, C.B. CAB International, Wallingford, UK, pp 345-364.

Staal, S. J., Owango, M. Muriuki, H., Kenyanjui, M., Lukuyu, B., Njoroge L., Njubi, D. Baltenweck I., Musembi F., Bwana O., Muriuki, K. Gichungu, G., Omore A., and Thorpe W. (2002a). *Dairy systems characterisation of the greater Nairobi milk shed*. Smallholder Dairy (Research and Development) project. Kenya Agricultural Research Institute (KARI)/Ministry of Agriculture (MoA)/International Livestock Research Institute (ILRI) Collaborative Dairy Research Programme, International Livestock Research Institute (ILRI), Nairobi, Kenya, 73 pp.

Staal, S. J., Baltenweck, I., Waithaka, M.M., deWolff, T. and Njoroge, L. (2002b). Location and uptake: integrated household and GIS analysis of technology adoption and land use, with application to smallholder dairy farms in Kenya. *Agricultural Economics* 27: 295-315.

Thorne, P.J. Thornton, P.K., Kruska, R.L., Reynolds, L., Waddington, S.R., Rutherford, A.S. and Odero, A.N. (2002). *Maize as food, fodder and fertiliser in intensifying crop-livestock systems in East and Southern Africa: an ex ante impact assessment of technology interventions to improve smallholder welfare*. ILRI Impact Assessment Series 11. ILRI (International Livestock Research Institute), Nairobi, Kenya, 123 pp.

12

Ecological impacts of sustainable systems for smallholders in Vietnam

B. Ogle[1] and T.R. Preston[2]

[1]*Department of Animal Nutrition and Management, Swedish University of Agricultural Sciences, Box 7024, 750 07 Uppsala, Sweden*

[2]*University of Tropical Agriculture Foundation, UTA-TOSOLY, AA # 48, Socorro, Santander, Colombia*

Introduction

The smallholder agricultural systems in Vietnam today, although still based largely on traditional systems and practices, are undergoing considerable change as a result of the economic and social reforms enacted under the process of "Doi Moi" and the resulting rapid economic growth of the 1990s. In particular paddy rice production has risen from an annual average of around 15 million tonnes for the period 1981-88 to over 20 million tonnes in 1999-2002, and this, and increases in the production of other major crops have been important factors behind the rise in both the numbers and productivity of the main livestock species, due partly to increased quantities of crop residues and agro-industrial by-products (e.g. rice bran) becoming available for feed. The rapid increase in Vietnam's Gross National Product (averaging 7.1% per annum between 1990 and 2000) has meant that consumers' disposable incomes have generally risen, and this has resulted in higher rates of consumption of animal products: between 1990 and 2000 the annual growth rate for total meat, milk and egg consumption (Table 12.1) was 5.4, 14.7 and 7.2%, respectively (FAOSTAT, 2003). However, much of this increase has come from higher inputs of grain based concentrates, which in turn implies increased inputs of fossil fuel energy, fertilizers, herbicides and insecticides etc., with associated negative impacts on both the local and global environments.

Integrated crop-livestock systems

In spite of the considerable advantages of integrated crop-livestock systems in Vietnam, and low-income countries generally, compared with specialized cropping or livestock production, it should be recognized that there are important socio-economic constraints to increasing animal production in mixed smallholder systems. In

benefits, including sales of ducks, were more than doubled. More importantly insecticide residue levels in rice in the control treatment were considered potentially hazardous for consumers, and were eliminated in the duckling-rice treatments. In a second experiment (Minh *et al.*, 2003) different fish species and densities, in combination with ducklings or layer ducks, were compared with a control treatment with conventional insecticide application. Rice yields were similar to the control treatment when fish alone were reared in the rice plots, but increased by 8.6 and 15.0%, respectively, when growing ducklings and layer ducks were included in the system, presumably because the scavenging ducks fertilize the growing rice with their droppings and stimulate development of the roots. Fish yields were more than doubled when ducks were allowed to scavenge in the rice plots compared to fish-rice, also as a result of the increased nutrients supplied by their droppings. The duck-rice-fish system thus gave very high economic benefits, over twenty times higher than for rice only, and more than four times higher than for ducks-rice and fish-rice. Similar studies were carried out in the Mekong Delta (Men and Ogle, 2003) to assess the ability of growing ducklings and fish to control weeds, insects and Golden Apple snails and to improve soil fertility in plots in which high-yielding rice varieties were cropped continually. Replacing herbicide and pesticide applications by growing ducklings and fish and reducing fertilizer application by 20% resulted in the almost complete elimination within two to three weeks of all major insect and weed pests and Golden Apple snails. Rice yields were generally higher in the integrated systems, the difference being significant for duck-rice-fish, particularly after three continuous crops of rice. This was a consequence of the more efficient removal of pests by the ducks and fish compared to pesticides, and also due to increased soil fertility compared to the control treatment, with significantly higher concentrations of nitrogen, phosphorus, potassium and organic matter (Table 12.2).

Table 12.2. Effect of ducks and fish on the soil fertility and soil organic matter content in rice plots (percentage).

	Control	*Ducks-rice*	*Ducks-rice-fish*	*P-value*
Nitrogen	0.207	0.250	0.286	0.001
Phosphorus	0.054	0.073	0.139	0.001
Potassium	1.74	2.16	2.67	0.019
Organic matter	6.89	7.07	7.88	0.054

Net profits for rice were 18 to 30% higher for the integrated systems compared to the conventional chemicals-based system, and between 55 and 144% higher if the profits from fish and duck sales were included. In traditional systems the duck flocks are also released in the rice fields directly after harvesting, where they consume leftover grains and weeds and fertilize the fields, thus also improving subsequent rice yields.

Agro-industrial by-products and crop residues

The economic and social reforms of the late 1980s resulted in vastly increased quantities of crop residues and agro-industrial by-products being made available in Vietnam. However, in many cases these are not fully utilized and instead become pollutants. For example, rice straw is often burnt, and shrimp by-products, cassava root pomace and brewery wastes dumped in rivers and water courses. A considerable amount of research has been carried out, therefore, over the last decade to develop and evaluate techniques for preserving and improving the nutritive value of crop residues and agro-industrial by-products of potential value as livestock feed.

Cassava by-products and residues

Cassava is widely grown in the more hilly areas of Vietnam unsuitable for rice cultivation and commercial production in the lowlands has increased in recent years to supply factories around Ho Chi Minh City and other cities producing starch, sodium glutamate and cassava root meal. The yield of cassava leaves at the time of harvesting the root is around 3 tonnes / ha, and crude protein content between 24 and 34% of dry matter (Phuc *et al.*, 2000). Different methods of processing the leaves (sun-drying and ensiling) were evaluated, both with respect to preservation and also reduction of the levels of toxins (HCN). Two days of sun-drying without pre-wilting reduced HCN concentrations in the fresh leaves by 90%, and six weeks of ensiling with molasses and salt by 62%. However, ensiling is the most attractive method of processing the leaves when the harvest coincides with the wet season, but is likely to be restricted to an "on-farm" operation due to the high volume and weight, and short shelf life after opening the silo. Sun-drying to produce a leaf meal has obvious advantages for commercial purposes. In central Vietnam the roots are harvested in the rainy season and have to be sold directly (and cheaply) to factories producing (expensive) cassava root meal. A technique was developed (Loc, 1996) to preserve and detoxify the fresh roots on farm by chopping and ensiling them in plastic sacks that were then covered with soil. The method was developed in cooperation with the local Women's Union and has proved to be very popular with smallholder pig producers in villages in central Vietnam.

Shrimp waste

In recent years the volume of processed shrimp products for export has increased by around 10% annually and in 1998 155,000 tonnes of shrimp were caught off the coast of Vietnam or cultured. Around 50% of the fresh weight of the shrimp consists of the head and shell (shrimp waste) which are removed during processing, and because of the difficulties of sun-drying the waste (a high proportion of the annual catch is taken in the rainy season) much is dumped by the processing factories into nearby rivers, where it constitutes a serious pollution hazard. The traditional method for preserving shrimp waste is by salt-based ensiling, but the high concentration of salt limits its use as an animal feed. Studies were carried by researchers from Hue University of Agriculture and Forestry (Ngoan *et al.*, 2000), therefore, to evaluate alternative ensiling techniques, and it was found that molasses and cassava root meal additives at ratios of 1:3 and 1:1 of additive to fresh shrimp waste, respectively, successfully preserved the waste. The resulting silage could replace 50% of the fish meal in diets for crossbred village pigs, and the increased usage by local producers has reduced the pollution problem in central Vietnam.

Rice straw

Vietnam currently produces an estimated 18 million tonnes of rice straw and much of this is burnt in the fields after harvest, with obvious negative environmental consequences. A series of trials has been carried out by the National Institute of Animal Husbandry in Hanoi and Cantho University, with Sida-SAREC support, on different techniques for improving the nutritive value of rice straw. These have confirmed, for example, that urea treatment and addition of molasses increase both intake and digestibility (Thu, 2003). The lack of grazing land and high price of ruminant meat have increased the interest in stall-feeding systems, and the supplementation of confined buffaloes and cattle with multi-nutrient blocks and cakes has been evaluated in a number of studies, with generally positive results (Do *et al.*, 2000). Although soil erosion caused by overgrazing is generally not a serious problem in Vietnam it does occur locally, and an expansion of confinement systems for ruminants for meat production based on treated or supplemented rice straw would alleviate the problem.

Brewery and distillers waste

By-products of industrial and artisan-scale alcohol production are produced on a daily basis in all the major cities towns and villages in Vietnam, which means that large quantities of brewery waste and rice distillers grains are available as animal feed, although they are often discarded, causing pollution. Experiments with growing meat ducks (Dong and Ogle, 2003) have shown that 50% of a commercial concentrate can be replaced by fresh brewery waste without any decrease in growth performance, and with higher net economic benefits for producers. Analyses and feeding trials with rice distillers waste carried out in the Mekong Delta (Manh *et al.*, 2003) indicated that the protein quality is comparable with that of the local fish meal, and that it can completely replace fish meal in diets for growing and fattening pigs. Local farmers are now using rice bran supplemented with distillers waste as the sole diet for their pigs.

Biodigesters

The introduction of the low-cost (< US$ 50 for a family size unit) plastic biodigester, constructed from tubular polyethylene film, put the technology within reach of a greater number of end users in Vietnam, where more than 20,000 households are estimated to be using the technology. Recent developments have focused on integrating the biodigester within the farming system and have demonstrated that the biodigestion process leads to major improvements in the value of the livestock manure as fertilizer for crops (Chau, 1998) and ponds growing water plants or fish (Sophin and Preston, 2002). As a result of the increasing emphasis on the promotion of farming systems based on the sustainable use of natural resources, it is now appreciated that the biodigester should be considered in a much wider perspective, and specifically in its potential role for the recycling of plant nutrients. This process has implications both as a means of reducing the dependence on inorganic fertilizers and for facilitating the production of foods and feeds of organic origin. Seen in this context, and in addition to being a renewable and sustainable form of energy, the biodigester plays several roles:

- It can be a means of de-contaminating wastes rich in organic matter. Concerns have been expressed in recent years on the effects of pollutants from animal wastes in relation to ecological and climatic changes. All animal waste products contain high levels of organic nutrients with the potential to decompose in the environment, with high biological oxygen demand (BOD), methane and ammonia emissions, and the release of excess nutrients and pathogens into ponds and waterways. In particular commercial and peri-urban pig producers are in a precarious position with the present system of waste disposal, and many units in residential areas in the cities of Vietnam have been closed by the authorities after complaints about odours from slurry discharged directly into small ponds and ditches. These odours are almost entirely removed if the slurry is passed through a biodigester, and in addition the digestion process results in the destruction of pathogens (Doelle, 1997). The decrease in the organic matter concentration allows the effluent to be used as a source of nutrients for fish production in ponds previously considered unsuitable due to eutrophication.
- Biogas can replace firewood as a source of fuel for cooking. The rapid rate of forest destruction and low rate of reforestation in

many countries in S E Asia have reduced the availability of firewood, and preserving forests by substituting biogas will have obvious environmental benefits, both locally and globally.

- The appropriate use of biodigesters can also give rise to a number of related socio-economic and health benefits, by eliminating the onerous and time-consuming task of collecting firewood by women and children, and by improving the kitchen environment through the elimination of smoke.

Multi-purpose trees

Small farmers in Vietnam and elsewhere are often reluctant to plant trees and shrubs to improve and protect soils from erosion, but they are more likely to be persuaded if it can be shown that the foliage is a valuable protein supplement for their animals. Promising trees and shrubs that have been studied as sources of protein for livestock in a wide range of ecosystems in Vietnam, are Jackfruit (*Artocarpus heterophyllus*) (Mui *et al.*, 2001) and *Gliricidia sepium* (Hao and Hiep, 2003) for goats, and *Trichanthera gigantea* for pigs and poultry (Nhan *et al.*, 1997). In particular feeding systems based on Jackfruit leaves have become popular among goat farmers in the hilly areas near Hanoi, and the excellent results found in terms of milk yield and growth rate (Mui *et al.*, 2001) have encouraged farmers to invest in confinement systems, thus reducing the risk of land degradation. *Trichanthera gigantea* has been shown to be a productive source of protein-rich biomass, and grows best under partial shade (Nhan *et al.*, 1996). It has been shown to adapt to a wide range of ecosystems in North (Binh, personal communication), Central (Ba and Ngoan, 2003) and South Vietnam (Nhan *et al.*, 1997).

High biomass crops

Sugar cane is known to be one of the most efficient crops for converting solar energy into multi-purpose biomass. However, its beneficial effects on soil fertility have not always been appreciated. Experiments in North Vietnam demonstrated the advantages of techniques for increasing total biomass yield and improving soil fertility, such as reducing planting distances and mulching with dead leaves, respectively (Mui *et al.*, 1996). Cassava is traditionally grown for root production, however, recent developments in Vietnam and Cambodia have shown that it can be managed as a semi-perennial forage crop with sustainable yields of more than 4 tonnes of protein / ha / year when heavily fertilized with goat manure or biodigester effluent (Preston, 2001). The sun-dried or ensiled leaves can be a partial replacement for fish and soya bean meals in pig diets (Phuc *et al.*, 2000).

Water plants

Traditionally water plants, both cultivated and wild, have formed an important part of the diets of humans and livestock in the Mekong Delta and other lowland areas in Vietnam and South East Asia. Studies have been carried out in Vietnam on a number of these, including for example water spinach (*Ipomoaea aquatica*), *azolla* and duckweed (*Lemna* spp). Water spinach grows widely in the Mekong delta, and is a highly productive source of protein-rich biomass, and like many other aquatic plants is fairly well digested by monogastrics and is low in anti-nutritional factors. Men (2000) showed that fresh water spinach could replace up to 30% of the concentrate in diets for dry sows, and improved reproductive performance. Duckweed in particular is a valuable source of nutrients for ducks (Men *et al.*, 1996) and pigs (Hang, 1998), with crude protein levels of up to 40% of dry matter when grown on ponds fertilized with the effluent from biogas digesters (Rodriguez and Preston, 1996). When managed correctly annual protein yields of over 7 tonnes / ha are possible (Khang, 2001). Duckweeds are also of particular interest due to their considerable potential to remove mineral contaminants from waste waters (Leng, 1999).

Conclusions

It seems inevitable that the recent introduction and spread of modern, large-scale industrial livestock units in Vietnam will continue, largely based on imported concentrate feeds and exotic breeds and hybrids, particularly for the production of broiler and pig meat for the tourist trade and export. The increased dependence on grains will have direct negative effects on both the local and global environment through increased inputs of fossil fuels and chemicals, and will also lead to problems for smallholder producers through competition for available feed resources and markets. This could result in a reduction in rural employment opportunities and increased migration to the cities. It could also lead to a decline in the numbers of indigenous pigs and poultry, which are the basis of traditional small and medium-scale poultry and pig meat production. It is important, therefore that integrated, environmentally friendly smallholder systems based on locally available feed resources continue to develop. Ways in which these systems can survive are through improvements in their productivity and through spreading an awareness of their advantages in terms of generating rural employment, providing the high quality, tasty products demanded by the discerning consumers of Vietnam, and not least by protecting the environment from the toxic residues inevitably resulting from industrial production systems.

Ngoan, L.D., Ogle, B. and Lindberg, J.E. (2000). Ensiling techniques for shrimp by-products and their nutritive value for pigs. *Asian-Australasian Journal of Animal Science* 13 (9):1278-1284.

Nhan, N.T.H., Hon, N.V., Son, V.V., Preston, T. R. and Dolberg, F. (1996). Effect of shade on biomass production and composition of the forage tree *Trichanthera gigantea*. *Livestock Research for Rural Development* 8 (2): http://www.cipav.org.co/lrrd/lrrd8/2/nahn.htm

Nhan, N.T.H., Preston, T. R. and Dolberg, F. (1997). Use of *Trichanthera gigantea* leaf meal and fresh leaves as livestock feed. *Livestock Research for Rural Development* 9 (1): http://www.cipav.org.co/lrrd/lrrd9/1/nhan91.htm

Phuc, B.H.N., Ogle, B. and Lindberg, J.E. (2000). Effect of replacing soybean protein with cassava leaf protein in cassava root meal based diets for growing pigs on digestibility and N retention. *Animal Feed Science and Technology* 83: 223-235.

Preston, T.R. (2001). Potential of cassava in integrated farming systems. *Proceedings of international workshop on current research and development on use of cassava as animal feed.* Edited by Preston, T.R., Ogle, R.B. and Wanapat, M. Khon Kaen University: http://www.mekarn.org/procKK/pres.htm

Rodríguez, L. and Preston, T. R. (1996). Use of effluent from low cost plastic biodigesters as fertilizer for duck weed ponds. *Livestock Research for Rural Development* 8 (2): http://cipav.org.co/lrrd/lrrd8/2/lylian2.htm

Sophin, P. and Preston, T.R. (2002). Effect of processing pig manure in a biodigester as fertilizer input for ponds growing fish in polyculture. *Proceedings of biodigester workshop, March 2002, Ho Chi Minh City.* http://www.mekarn.org/procbiod/pich.htm

Thu, N.V. (2003). *In vitro* digestibility of rice straw after mushroom cultivation and the utilization of nutrients of enriched rice straw by Red Sindhi crossbred cattle. In: *Proceedings of final national seminar-workshop on sustainable livestock production on local feed resources.* Edited by Preston, T.R. and Ogle, R.B. HUAF-SAREC, Hue City, 25-28 March, 2003. http://www.mekarn.org/sarec03/thu1.htm

Xuan, V.T. (1999). *Asian ecotechnology workshop and fieldwork programme for Southern Vietnam*. Ho Chi Minh City and Cantho City, Vietnam.

13

Research to alleviate poverty in the face of industrialisation of livestock production

Ann Waters-Bayer[1] and Wolfgang Bayer[2]
[1] *ETC Ecoculture, POB 64, NL-3830 AB Leusden, Netherlands*
[2] *Independent advisor, Rohnsweg 56, D-37085 Göttingen, Germany*

Introduction

Economic projections into the year 2020 (Delgado *et al.*, 1999) indicate that the global demand for animal products will increase sharply, particularly in developing countries, and suggest that the bulk of the increased production to meet this demand will come from large-scale, industrial production. Many smallholders in developing countries depend at least partly on livestock as a source of livelihood; for pastoralists, livestock are by definition the main source of livelihood. For the poorer livestock-keepers, particularly the landless, the animals are often more vital for income and security than they are for better-off producers. The forecasts indicate that – if appropriate policies are not introduced – the industrialisation of livestock production will have serious negative social and economic consequences for these livestock-keepers in developing countries. In the face of this, what can and should be the role of animal science research?

In this paper, we examine the past role of animal science research in developing livestock production in the "North" and in the "South", highlighting how research has served a particular type of production system. We then look at what animal science research has contributed to smallholder, pastoral and landless livestock-keeping systems. Finally, we put the focus on public-sector animal science research committed to poverty alleviation, and the new types of partnership in research and development that can benefit the poorer people involved in livestock-keeping.

The role of animal science research in developing livestock production

Animal science research in the "North"

Animal science research started in industrialised countries about 200 years ago. Many of the early agricultural (including animal) scientists were gentleman farmers who owned or managed large estates (Hippauf, 2000). Primary attention, especially in veterinary science and breeding, was given to horses on account of their importance for upper-class pleasure, representation and warfare. Otherwise, for productive purposes, the main species with which animal science dealt were cattle (for beef, milk and, to a lesser extent, draught), sheep (for wool), pigs (for lard and meat) and, at a later stage, chickens (for eggs and meat). In terms of topics, the focus was on animal health, breeding and nutrition and forage production. The main objectives of the research were to achieve higher production of all above-mentioned products and higher quality, specifically, of wool.

Agricultural colleges, the earliest of which were established in the late 18th century in Europe, conducted research and teaching not primarily on the art of managing the small farms from which most of the rural population in Europe depended for their livelihoods, but rather on the management of the large estates. The new technologies developed in breeding, stabling, feeding and milking were designed for medium- to large-scale farms, and not for smallholdings (Franz, 1968). A similar trend could be observed in the United States with the land-grant colleges and in Australia and New Zealand, where research and teaching in animal science focused on specialised livestock production units of an economically viable size.

Research programmes were designed in close cooperation between researchers and the State, societies for the advancement of agriculture and farmer organisations in which large-scale producers were particularly influential (Middendorf and Busch, 1997). In many countries, there is also fairly close cooperation between research institutes, universities and extension services, especially in the land-grant colleges in the USA. Professional associations such as the national, continental or international societies for animal breeding or for animal nutrition involve staff from research, teaching and extension.

Over time, other actors in the livestock industry became customers of animal science research: the service and input suppliers. Attention was given to the design of ploughs and other animal-drawn farm implements, mainly for use with horses and oxen. Feed manufacturers benefited from feed analysis systems developed by animal scientists. Breeding associations benefited from breeding plans and scientific criteria for selection, such as the size of a certain muscle. Food processors benefited from the development of new products and equipment to produce them. As we all know, the primary aim – to increase animal productivity – was achieved with splendour. Meat, eggs and milk became cheaper and cheaper. In most of the technologies that emanated from research, economies of scale could be achieved. As a result, primarily the larger-scale enterprises reaped the benefits, whereas small-scale producers benefited more by chance than by intention. The social change during industrialisation in the 19th and 20th centuries lead to a reduction in the number of farmers and to an increase in farm size, and the importance of agriculture for the overall economy in industrialised countries declined.

Thus, the type of animal science research conducted over the past two centuries has been closely linked with industrial development and has served a particular type of production system: specialised, commercialised and male-dominated.

Likewise with a view to increasing production, the environment in which the animals live has been adapted to the needs of the animals, rather than seeking animals adapted to the environment. In modern livestock enterprises, the temperature, air humidity and light in animal housing is regulated, at considerable expense in terms of fossil fuels. Feed and forage management is based on a nutritional "demand system": animal requirements for high production are determined, the potential intake is estimated and the rations are formulated accordingly. Previously, livestock had to adapt themselves largely to "supply", i.e. to the availability of feed and water, the quantity and quality of which fluctuated greatly from season to season. They had to rely on mechanisms such as thicker coats or fur to protect themselves against heat and cold. Strict control of the animals' environment is easier and less costly per animal in enterprises with a large number of animals than it is in smaller animal holdings. Also new technologies such as computerised calculation of rations and computerised feed mixtures are hardly suitable for a farm with only 3–4 cows.

In pursuit of increased production per animal, better-quality feed is needed in larger quantities than can be furnished by grassland, crop residues or forests. At present, the greater part of available grain in the USA and Europe is not used for human nutrition but as livestock feed. Livestock keeping used to be complementary to the production of food crops but, in specialised and industrialised systems, it has become a competitor for food that could also be consumed directly by humans. Grain is fed not only to monogastric animals but also to ruminants such as beef and dairy cattle and to sheep for the production of "fat lambs". Manure, which farmers used to regard as an asset for maintaining soil fertility, has become a liability in some areas, because there is not enough land close to the large animal holdings to absorb all the manure produced. In the name of increased productivity, livestock growth promoters and hormones were and, to some extent, are still used, whereby the possible effects on humans are far from clear.

These industrialised systems have been increasingly criticised, and animal science research in the North has been obliged to react. Issues of environmental impact have become more prominent in international negotiations, landscape maintenance has become a task for livestock-keepers and issues of animal welfare are of concern to the general public. Particularly in the last couple of years in Western Europe, food safety has become a burning issue. The changing priorities in the public and political spheres are reflected in a re-organisation of government ministries, research organisations and, to a lesser degree, universities. In some states in Germany, agriculture is now administered by the Ministry for Environment and, on federal level, it is combined with consumer protection.

Animal science research in the "South"

Animal science research and teaching in developing countries have been strongly influenced by concepts and production systems in the industrialised countries. Pastoral and small-scale livestock-keeping was regarded as backward, unproductive and needing to be replaced. However, after many decades of research and development focused on "modern" livestock production, millions of people in developing countries are still practising indigenous forms of livestock-keeping not only in rural but also in peri-urban and urban areas. Livestock development in Africa for the period 1950–80 was described as "thirty years without progress" (Jahnke, 1982). Almost two decades later, a review of livestock development projects by major donors (LID, 1998) revealed little evidence of sustainable impact on the livelihoods of the poor. Animal science research "contributed" its share to this failure, as it neglected the real-life production systems of the vast majority of livestock-keepers in the developing countries. Most researchers have given little attention to how improvements can be made in the indigenous and, especially in the case of urban livestock-keeping, the spontaneously emerging systems. Governments in both the South and the North have largely disregarded the impact of national and international policies on the livelihoods and welfare of the people who depend on these indigenous and very widespread forms of livestock-keeping.

In the last two decades of the 20th century, there has been some gradual change in animal science in the South. Livestock husbandry by peasant farmers and pastoralists in developing countries had initially been investigated mainly by social and anthropological researchers, but these were joined by some animal scientists who took a farming/livestock systems perspective. This research revealed that "traditional" animal husbandry as practised by smallholders and pastoralists is not less rational than "modern" animal production, but the objectives differ. Whereas "modern" animal production is aimed at supplying markets with food and fibre, smallholder and pastoral animal husbandry pursues diverse goals including production for both household and market, building up capital, maintaining savings, buffering against risk and fulfilling social and cultural functions. The products include not only food and fibre, but also draught power, transport capacity and manure to be used as fertiliser or fuel while the animals continue to live, if only at survival level, and various products from the eventually slaughtered animal such as skins, bones and horns. Moreover, the traditional systems have not been static. Local experimentation and innovation by livestock-keepers and suppliers of inputs such as feed in the informal sector have led to changes that most animal scientists have overlooked. The system studies often did not take these local dynamics into account.

Contribution of animal science research to smallholder, pastoral and landless livestock-keeping

To be sure, many of the system studies did lead to a more differentiated view of livestock-keeping and a greater understanding of the importance of livestock for rural livelihoods. However, the "hard" component research work in animal science – and indeed many of the recommendations for improving the livestock systems that have followed on these studies – tend to revert back to the standard solutions that served high-production-oriented and specialised livestock enterprises. The disciplinary work within animal science, which is so necessary for optimising resource-limited production, has been adjusting only very slowly to the new insights being gained by system studies of livestock-keeping by small-scale farmers, pastoralists and the landless. The component research often relies on methods and measurements that were developed for different conditions and levels of production – i.e. on the controlled and high-yielding end of the scale rather than on the more natural and multi-functional end of the scale.

Productivity studies

It was in the late 1970s and early 1980s that animal scientists started to look more closely at the productivity of exotic and indigenous breeds. The research covered primarily cattle, sheep and goats. The weight of weaned offspring per dam was regarded as the best indicator for herd productivity. Even from this limited perspective, breed comparisons on station showed that – already under conditions of moderate stress – the performance of indigenous breeds with respect to meat production could almost match that of exotic breeds. When the weight differences of the dams were taken into account and weaning weight per 100 kg of dam or per kg metabolic weight was calculated, indigenous breeds were similar to or better in their performance than exotic breeds (e.g. Buck *et al.*, 1982).

In systems with meat production as the primary aim, weaning weight is important but how relevant is this if animals are multifunctional? Ridder and Wagenaar (1986) addressed some of these issues when they compared the productivity of ranches with that of communal animal husbandry in Botswana. Whereas ranches produce beef only, communal systems produce also milk and draught power. The outcome of this comparison was that, when all major products were taken into account, the productivity per animal was similar in both systems and, on a per hectare basis, the traditional (agro)pastoral system out-yielded the modern ranching system. However, this comparison still did not give a value to other functions of livestock such risk aversion or social security. Workneh Ayalew (2000), comparing crossbred and local goats in Ethiopia, included not only meat, milk and manure but also quantifiable socio-economic data, i.e. financial and insurance (security) benefits. This study revealed that, although crossbred goats gave more milk and grew faster, they also lost weight faster than did the indigenous goats during the lean season and were therefore more at risk and had a lower "insurance value".

In order to achieve a more realistic assessment of livestock productivity, the productivity studies need to be conducted under field conditions in smallholder and pastoral systems but, mindful of the accuracy required for publication in double-refereed journals, such productivity studies must be long-term. It is difficult to ensure the cooperation of livestock-keepers for these studies, if they can see no short-term benefits. Moreover, assuring the collection of good-quality scientific data under field conditions proved to be tedious and expensive. This work was done mainly within the framework of development projects and applied research financed out of development funds, and the results were meant to help in designing better projects. Under such circumstances, waiting five years before a baseline study on animal production was carried out, analysed and published was simply unacceptable. Already in the early 1980s, short-cut methods were sought and attempts were made to "interview cows", relying on herders' and animal managers' memories. Initially called "quick and dirty", such methods became more respectable under the name of Rapid Rural Appraisal (RRA) and Participatory Rural Appraisal (PRA). Their application under smallholder conditions was useful in revealing the main bottlenecks in animal production much more quickly.

In short, various methods have been applied to assess productivity in smallholder and pastoral livestock-keeping systems, and more differentiated approaches to measuring animal productivity are gradually being developed. Those which seek to measure contribution of livestock to livelihoods have given credence to the "conservative" approach of many smallholders and pastoralists with regard to adopting so-called improved breeds and are laying the foundation for more promising approaches in livestock research and development for poverty alleviation.

Moreover, the results of this research suggest that, if the multi-functionality of animal husbandry is taken into account, the Livestock 2020 study (Delgado *et al.*, 1999) underestimates the present overall benefits of animal husbandry and also underestimates the losses that will be incurred if the existing multi-functional systems are replaced by industrialised systems of livestock production.

Animal nutrition studies

In animal nutrition studies, the situation does not differ much from livestock productivity studies. The approach and methods were likewise derived from a science designed for specialised, high-output production systems not relevant for smallholders and pastoralists. Initially, much emphasis was placed on conventional methods of evaluating the nutritive value of feed and forage for animal growth, lactation and egg production. The bulk of this nutritional research concentrates on only a few species: cattle, pigs, chickens and, more recently, some fish species such as salmon and trout. These are species of great interest to industrialised production. Already when it comes to sheep, nutrition has been less researched, even though the digestibility of feed and forage can differ greatly between cattle and sheep. In the case of other livestock species kept by smallholders and pastoralists, such as buffaloes, goats, donkeys or camels, the situation is even worse. A book on donkey nutrition, for example, is very hard to come by, despite the fact that development practitioners have recognised since decades the great importance of this so-called "minor species" for smallholders and pastoralists – especially to women to relieve them of some of their many burdens. With respect to nutritive value of feed, animal scientists know a great deal about digestible protein and mineral content of some "improved" forage species, such as *Panicum maximum* or various *Stylosanthes* species, but women farm labourers in India doubtless know more about the nutritive value of different types of weeds for different species and classes of livestock (Rangnekar, 1998) than do most scientists.

Calculations of the feeding value of forages are still often based on an oversimplified concept that animals eat whatever is put in front of them. They are still often determined on the basis of whole-plant analyses or, in the case of pastures, cut samples. However, grazing animals are selective feeders and select between plant species and between genotypes on the species level. They generally prefer leaves to stem and green over dry matter. In pasture science, it is common that cut samples suggest a digestible nutrient content that is below the maintenance requirements of the animals, yet the animals continue to gain weight because they select only the best part of the forage. Much of the conventional work on animal nutrition has shown how precise science is measuring precisely the wrong thing.

These feed evaluation methods are of high value for industrialised livestock production, where precise knowledge of nutritional requirements for production is important and feed rations can be adapted to the animals' needs, increasingly accurately with the help of computers. Under the low-external-input conductions in which the poorer livestock-keepers have to work, forages and feeds are finite resources, and supplements and concentrate feeds are not available or not affordable. Therefore, an approach is needed that optimises the use of available forage resources rather than maximises the production per animal. But even optimisation is not an absolute – it depends on the production aim(s). Pasture science work undertaken already almost 30 years ago found a substantial difference between optimal stocking rate for maximum gain per animal and that for maximum gain per unit area (Jones and Sandland, 1974). These results revealed that the optimal stocking rate on a pasture would depend on the objectives of the livestock-keeper. For example, if the main aim is to produce young fat animals or to produce milk, i.e. production traits for which livestock need good nutrition, a low stocking rate is advantageous. If the main aim is to keep livestock as a savings account or to produce manure, the stocking rate can be high, as the animals need to survive but need not grow quickly. If a livestock-keeper has multiple objectives with different nutritional requirements, the optimal stocking rate or feeding level will be somewhere between these two extremes.

Researchers working in Indonesia applied this concept of optimal feeding level also to stall-feeding by smallholders. In a regional study in Eastern Java, the available quantities and qualities of the different types of forages were estimated or measured. The optimal forage utilisation for overall liveweight gain was reached when the best 35% percent of the available forage was used. At this level of utilisation, the gain per animal was about half of the gain per animal using the best 10% of the forage. When manure was given a value (30 kg of liveweight gain per year) and the model was re-run with manure as a product, the optimal level of forage utilisation was reached when the best 70% of the forage was used. At this level of utilisation, the liveweight gain per animal was only about a quarter of what could be achieved if only the 10% best-quality forage was used (Zemmelink, 1986; Zemmelink *et al.*, 1992; Bayer and Zemmelink, 1998). The fact that most of us would regard these animals as poor skinny beasts that need more meat on their ribs indicates how little multi-functionality and its implications are embedded in our minds.

Breeds, biodiversity and animal adaptation to the environment

Also with regard to managing animal genetic resources, many researchers and research planners need to re-think. As a first approximation, livestock biodiversity is measured in terms of the number of breeds, and biodiversity is endangered when breeds become extinct or are endangered. FAO (2000) prepares and updates a World Watch List. However, a closer look reveals that the definition of breeds is far from clear. Breeds in the "North" are associated with herd-books with narrow breeding objectives and ideal animals. The focus on keeping particular standardised breeds can reduce rather than enhance biodiversity. Smallholders and pastoralists do not need particular breeds but rather animals with particular characteristics that suit specific environments and production systems, characteristics such as ability to cope with low feed quality or little water or, in the case of mobile pastoralism, ease of herding and ability to walk long distances. Thus far, the scientific investigation of many of these traits is anecdotal rather than systematic. Furthermore, as has been found by Adams *et al.* (2002), pastoralists may want several types of animal in their herds to be able to benefit when the natural conditions are good but, at the same time, to have some buffer against risks such as drought or disease. They do not want herds composed of highly uniform and supposedly "ideal" animals suitable for only one set of conditions.

Another area to which more attention needs to be given is animal adaptation to the environment. There was a period, some 40–50 years ago, when such research was popular in the USA, South Africa and Australia, when it became apparent that the environment in most of those areas could not be adapted in an economic way to the needs of high-yielding breeds. It could be shown that local stock or – in the case of Australia – animals coming from Asian stock (*Bos indicus*) are more tolerant to heat and have lower maintenance requirements than the stock imported from temperate climates (in the case of cattle, *Bos taurus*). Animals adapted to hot and arid environments have low water turnover (King, 1983) and, in many cases, indigenous animals are resistant to or can tolerate diseases that are deadly to imported stock (e.g. Bayer and Maina, 1984). However, these adaptive traits are antagonists of the productive traits of higher milk yield or growth rates. During periods of stress, the "biggest calves die first". Perhaps with the exception of disease resistance, research on animal adaptation to the environment is low-key at present. On the contrary, many current breeding programmes still emphasise the need to select for faster growth or higher milk yield, without considering the changes in the environment that are prerequisite for the expression of such traits, and the dependencies which will be created among livestock-keepers if the conditions are temporarily changed, e.g. through project-subsidised supply of better-quality feed. With a view to sustainability and security of livestock production by poor smallholders and pastoralists, research on animal adaptation should have a much higher priority, at the very least to underline the importance of considering adaptive traits in livestock development plans.

It is indeed good to see the attention given in recent years to so-called minor or non-conventional livestock species such as guinea pigs (e.g. BOSTID, 1991; Valle-Zárate, 1999), yaks (Li and Wiener, 1995), llamas and alpacas (Fernandez-Baca, 1990). Care must be taken that the role of these animals is understood within the framework of the respective production systems and that research and breeding efforts are not channelled primarily towards producing super fast-growing guinea pigs and super high milk yielders among camels, buffaloes or yaks, as the very advantages of these species and breeds for poor livestock-keepers living in uncertain environments may be lost in the process. Of greater interest, for the sake of more secure livelihoods and greater agricultural biodiversity, would be to explore ways of better integrating the keeping of a variety of species for different niches, including co-management of domesticated and wild or semi-domesticated species.

Animal science research to alleviate poverty

At first glance, the official agendas of many international and national public-sector research institutes suggest that things are looking up. So-called "pro-poor" strategies can now be read almost everywhere and heard at almost every international conference. However, we still face a major challenge because, on the operational level, the same old conventional methods, which are inappropriate for "pro-poor" research, often still prevail. We have to critically examine our toolbox and may have to throw some tools out, modify others or acquire new ones. We also have to differentiate between precision and relevance. For example, forage-use calendars compiled together with landless livestock-keepers are clearly rough and tough and approximate, but they may be much more relevant for identifying the main problems that these livestock-keepers face than the precise chemical analysis of plants or plant parts.

Some of the examples given above – and some papers presented at this conference – do bear witness to a gradual change in research questions, approaches and methodology so that animal science research can contribute to poverty alleviation. However, most conventionally trained animal scientists still do not seem to have recognised a) that their animal science – which many of them regard as "universal" – is aimed primarily at high-external-input and large-scale enterprises; and b) that smallholder and pastoral systems of managing livestock and land is not just a scaled-down version of large commercial farms. Research questions and methods need to take these differences into account.

An approach to animal science research that aims to contribute to alleviation of poverty among smallholders and pastoralists must have a different agenda from conventional science for animal production. Some of the many issues that deserve attention in a "pro-poor" approach to animal science research include:

- development of site-appropriate low-external-input technologies in livestock health, feeding and husbandry that can be afforded by poor people
- identifying – and, if possible, making small improvements in – animal genotypes adapted to the environment, such as to disease pressure, high ambient temperatures, water scarcity and low-quality feed
- seeking ways to optimise the use of available feed resources for different purposes rather than maximising production with external resources
- exploring the potentials of "non-conventional" animals, especially those that can be kept with little land and other resources, and semi-domesticated wildlife that are harvested by poorer households
- exploring ways of improving nutrient cycling not necessarily within production units, as in the typical model for mixed farming, but rather between different types of production units, e.g. crop farming and pastoral, and/or between production and consumption units, e.g. in urban and peri-urban areas.

Even the more recent attention to crop-livestock integration within mixed-farming systems – although an improvement over the focus on specialised livestock systems – is still not addressing directly the concerns of the poorer livestock-keepers or those who could be helped out of poverty through livestock-keeping. The model of mixed farming on which these approaches are based is a model of an intact family that has access to both land and livestock so that it can integrate production and can focus on this crop-livestock integration. Such families are among the relatively well-off ones in rural areas (Scoones and Wolmer, 2002). The poor include the women-headed households, the households weakened by illness or disease, the households without secure rights to land or water, the households that are obliged to seek multiple sources of livelihood and cannot concentrate on their livestock or perhaps even their crops, if they have any access to land at all. Poor families are obliged to follow different paths than better-off smallholders in order to secure, let alone to improve their livelihoods. The roles of livestock are different in the poorer households and the ways in which livestock-keeping can contribute more to livelihood security will likewise differ. Livestock-related research aimed at alleviating poverty must seek a better understanding of these paths in order to be able to accompany the livestock-keepers to improvement, and to help policymakers and development planners understand them.

Research focused on poverty alleviation needs to go beyond biophysical innovations in livestock production, to include issues of:

- access to rights to use water and land resources for livestock-keeping;
- access to forms of credit and mutual insurance systems that allow poor people, without own capital, to obtain animals and enter into small-scale enterprises at low risk;
- approaches and methods to support the process of building local institutions that give marginalised people – particularly women – more control over production, processing and marketing and increase their capacities to operate their own institutions, such as through self-managed decentralised animal health care or small-scale processing units to add value to livestock products; and
- socio-economic and ecological impacts of livestock-related policies.

There is a need for refocusing not only in the topics of research but also in the approach to research, towards participatory research that stimulates local experimentation and innovation by men and women in poorer households in order to develop site-appropriate technologies and local institutions for animal production, processing and local and regional marketing. The conditions under which they live are diverse and changes are frequent – also in terms of market accessibility for those animals and products they choose to sell. Their survival depends on their capacities to adjust and also to diversify into activities in addition to livestock-keeping. An approach to research and development that strengthens local capacities to innovate and adapt will, in itself, contribute to human resource development. Likewise, the policy research related to livestock should be participatory and action-oriented so that it gives small-scale producers, local processors and consumers a stronger voice in promoting production systems with positive ecological and socio-economic effects and in reducing the market-distorting subsidies for and negative environmental impacts of industrialised livestock production.

This call for participatory research does not deny the importance of more conventional disciplinary research under controlled conditions – on research stations, in laboratories etc. But there is a need for closer linkages between this component research and system-based research rooted in the practice of smallholders and pastoralists and for a re-thinking within the components so that the disciplinary research can be more sharply focused on issues relevant for practice and relevant for the poor.

New partnerships in livestock-related research to alleviate poverty

These different topics and approaches demand that "pure" animal scientists link up much more closely with other actors in developing improved rural and urban livelihoods, transcending the conventional biophysical focus. Cutting-edge science will be science that addresses the challenge of redesigning research for poverty alleviation – and there are precious few scientists close to this cutting edge at the moment.

Large-scale commercial livestock systems oriented toward high levels of production will continue to be supported by animal research in the private sector. In relation to this, the public-sector research will need to look at environmental and economic policies that do not give unfair advantage to large-scale high-external-input producers and that require them to carry the full costs of the environmental damage they cause. In both cases – whether in cutting-edge livestock research for and with poor rural and urban producers or in research to lay the foundations for a level "playing field" with commercial enterprises – researchers from animal science will increasingly have to enter into new types of partnership with different disciplines and with different actors in research and development.

Much of the research to develop technologies relevant for the poor requires close collaboration between scientists, development agents, livestock-keepers, suppliers of appropriate inputs and processors and buyers of livestock products. Together they will need to analyse the situation, identify priority problems and opportunities for improvement, and explore these opportunities together through joint experimentation, monitoring and evaluation. Also the research related to institutions that can improve poor people's access to the resources needed for livestock production requires collaboration in similar types of processes with local administrations and policymakers at various levels from local to international.

Such development-oriented participatory research erases the artificial boundary between research and "extension". It is not a question of developing solutions in isolation, then passing these on to extension agents so that the technologies can be transferred to the users. The innovations are developed together with producers who are themselves seeking ways to improve their situation. The research and experimentation involving livestock-keepers and various other stakeholders and the sharing of results among them leads to the dissemination of new ideas and stimulates further research and experimentation by both producers and formal scientists. The livestock-keepers involved in the research can also make important contributions to developing appropriate dissemination strategies.

Over recent years, some cases of such participatory research and development have been realised and documented, such as the Forage for Smallholders Project in Southeast Asia (Horne *et al.*, 2000), the development of donkey-drawn ploughs in collaboration with local blacksmiths in Sudan (Majzoub, 2001), the support to joint experiments by crop farmers and mobile pastoralists to optimise overnight kraaling of animals for production of local cash crops in Cameroon (Tchawa, 2000) and some examples bought in this conference. However, these promising examples have been on the margins of a mass of conventional animal science research that still has little to do with the specific characteristics of small-scale and pastoral livestock-keeping systems.

Meeting the challenge

In order to bring about this revolution in animal science research so that the cutting edge can be reached and the challenge of reducing poverty can be addressed, changes are needed in the education of animal scientists and in the criteria according to which excellence in science is assessed. College and university education must include much more direct exposure to poor livestock-keepers – not simply excursions to large-scale commercial dairies and chicken factories on the edge of town, but rather studies of innovations and coping mechanisms in poorer rural and urban livestock-keeping households and groups. Disciplinary teaching should be set within the framework of livelihoods systems, and build the capacities of students to engage in participatory situation analysis and to transform findings into relevant research questions in breeding, nutrition, husbandry practices, processing etc. It should teach students how to engage in real dialogue with livestock-keepers who may have little formal education but have a wealth of indigenous knowledge that needs to be recognised. It should give students ample opportunity to carry out research directly with small-scale livestock keepers and pastoralists, as well as local producers of livestock inputs and equipment and, also when the research is station-based, to interact directly with livestock-keepers in assessing the results and their relevance.

Greater recognition needs to be given to scientists who engage in research that makes a contribution to improving the livelihoods of the poor, e.g., through offering publishing and learning opportunities, through giving awards and promotion. The heads of research institutions are key figures in creating a professional culture of science serving development and in guiding scientists to a new self-image of their role for the public good. Other important figures that will play a major role in bringing about this change are the institutions that finance research and that issue calls for proposals; the members of committees who select candidates for positions in research, administration and universities; the referees for scientific journals; and the evaluators of research proposals and projects. All these should be re-examining the criteria they use and making the necessary changes to render animal science research more relevant for the poor.

International conventions related to conservation and biodiversity management are opening up new doors for cutting-edge research that recognises multiple functions of resources, including domestic livestock and wildlife, within livelihood systems. These conventions create favourable conditions for new emphases in research – both new sources of funding and new sources of recognition that transcend the narrow earlier focus in animal science research on high levels of production, regardless of environmental and social costs. This may provide new and additional incentives for animal scientists, beyond the satisfaction of carrying out research that has a positive impact on the poor. They offer openings for imaginative, innovative and committed researchers who recognise the potentials of livestock for the livelihoods of the poor.

References

Adams, M., Kaufmann, B. and Valle-Zárate, A. 2002. Indigenous characterization of local camel populations and breeding methods of pastoralists in Northern Kenya. Paper presented at Deutscher Tropentag 2002, Witzenhausen, 9–11 October 2002.

Bayer, W. and Maina, J. 1984. Seasonal pattern of tick load in Bunaji cattle in the subhumid zone of Nigeria. *Veterinary Parasitology* 15: 301–307.

Bayer, W. and Zemmelink, G. 1998. Management of livestock and forage resources. *Forage husbandry* (W. Bayer and A. Waters-Bayer), pp. 41–65. Macmillan Education Ltd, London, UK.

BOSTID (Board on Science and Technology for International Development, Office of International Affairs, National Research Council). 1991. *Microlivestock: little-known small animals with a promising economic future.* National Academy Press, Washington DC, USA, 449pp.

Buck, N., Light, D., Lethola, L., Rennie, T., Mlambo, N. and Muke, B. 1982. Beef cattle breeding systems in Botswana – the use of indigenous breeds. *World Animal Review* 43: 12–16.

Delgado, C., Rosegrant, M., Steinfeld, H., Ehui, S. and Courbois, C. 1999. *Livestock to 2020: the next food revolution*. Food, Agriculture, and the Environment Discussion Paper 28. IFPRI, Washington DC, USA; FAO, Rome, Italy; ILRI, Nairobi, Kenya, 72pp.

FAO. 2000. *World Watch List for domestic animal diversity*. 3rd edition. FAO, Rome, Italy, 726pp.

Fernandez-Baca, S. 1990. Llamoids or New World Camelidae: llama, alpaca, guanaco and vicuna. In: *An introduction to animal husbandry in the tropics* (W.J.A. Payne). 4th ed. Longman, Harlow, UK, pp 557–580.

Franz, G. (ed.) 1968. *Universität Hohenheim: landwirtschaftliche Hochschule 1818–1968*. Verlag Eugen Ulmer, Stuttgart, Germany, 331pp.

Hippauf, R. 2000. *Johann Heinrich von Thünen*. Hinstorff Verlag, Rostock, Germany, 221pp.

Horne, P.M., Magboo, E., Kerridge, P.C., Tuhulele, M., Phimphachanhvongsod, V., Gabunada Jr., F., Le Hoa Binh and Stür, W.W. 2000. Participatory approaches to forage technology development with smallholders in Southeast Asia. *Working with farmers: the key to adoption of forage technologies* (ed. W.W. Stür, P.M. Horne, J.B. Hacker and J.C. Kerridge), pp. 23–31. Australian Centre for International Agricultural Research, Canberra, Australia.

Jahnke, H.E. 1982. *Livestock production systems and livestock development in tropical Africa*. Wissenschaftsverlag Vauk, Kiel, Germany, 253pp.

Jones, R.J. and Sandland, R.L. 1974. The relation between animal gain and stocking rate: derivation of the relation from the results of grazing trials. *Journal of Agricultural Science (Cambridge)* 83: 335–341.

King, J.M. 1983. *Livestock water needs in pastoral Africa in relation to climate and forage*. Research Report 7. International Livestock Centre for Africa, Addis Ababa, Ethiopia, 95pp.

Li, C. and Wiener, G. 1995. *The yak*. FAO RAP Publication No. 1955/25. Regional Office for Asia and the Pacific of the Food and Agricultural Organization of the united Nations, Bangkok, Thailand, 237pp.

LID (Livestock in Development). 1998. Strategies for improving DFID's impact on poverty reduction: a review of best practice in the livestock sector. LID, Crewkerne, UK, 48 pp + annexes.

Majzoub, M. 2001. The experience of ITDG in participatory development of the donkey-drawn plough in North Darfur, Western Sudan. Paper presented at International Workshop on

Advancing Participatory Technology Development, 17–21 September 2001, International Institute for Rural Reconstruction, Silang, Cavite, Philippines.

Middendorf, G. and Busch, L. 1997. Inquiry for the public good: democratic participation in agricultural research. *Agriculture and Human Values* 14: 45–57.

Rangnekar, S. 1998. Expert animal nutritionists. *Forage husbandry* (W. Bayer and A. Waters-Bayer), p. 151. Macmillan Education Ltd, London, UK, 198pp.

Ridder, N. de and Wagenaar, K.T. 1986. Energy and protein balances in traditional livestock systems and ranching in Botswana. *Agricultural Systems* 20: 1–16.

Scoones, I. and Wolmer, W. 2002. *Pathways of change in Africa: crops, livestock & livelihoods in Mali, Ethiopia & Zimbabwe.* James Currey, Oxford, UK / Heinemann, Portsmouth, USA, 236pp.

Tchawa, P. 2000. Chains of innovation by farmers in Cameroon. *LEISA Magazine* 16 (2): 14–15.

Valle-Zárate, A. 1999. Meerschweinchen. *Tierzucht in den Tropen und Subtropen.* Handbuch der Landwirtschaft und Ernährung in den Entwicklungsländern, Vol. 5 (ed. P. Horst and I. Reh), pp. 353–364. Ulmer Verlag, Stuttgart, Germany.

Workneh Ayalew. 2000. *Do smallholder farmers benefit more from crossbred (Somali x Anglo-Nubian than from indigenous goats?* Cuvillier Verlag, Göttingen, Germany, 155pp.

Zemmelink, G. 1986. Measuring intake of tropical forages. *Recent advances in feed rationing systems for dairy cattle in extensive and intensive production systems* (ed. C.C. Balch and A.J.H. van Es), *International Dairy Federation Bulletin* 196: 17–21.

Zemmelink, G., Brouwer, B.O. and Subagiyo, I. 1992. Feed utilisation and the role of ruminants in farming systems. *Livestock and feed development in the tropics* (ed. M.N.M. Ibrahim, R. de Jong, J. van Bruchem and H. Purnomo), pp. 444–452. Brawijaya University, Malang, Indonesia.

14

Meeting the increased demand for animal products in Asia: opportunities and challenges for research

C. Devendra
(International Livestock Research Institute, PO Box 30709, Nairobi, Kenya). Contact address: 130A Jalan Awan Jawa, 58200 Kuala Lumpur, Malaysia

Introduction

Improved animal production and productivity enhancement in Asia are justified in direct response to the need for more animal proteins. This need is associated with rising incomes, which encourage people to diversify their diets to include a variety of meats, eggs and dairy products, including the substitution of calories in livestock for low-priced starch calories. The projected total meat and milk human consumption levels in 2020 are far in excess of anticipated supplies (Degaldo *et al.*, 1999). The rapid growth in the consumption of foods of animal origin is especially spectacular in East and South East Asia where the demand growth has placed unprecedented pressure on the management of the natural resources (crops, animals, land and water).

The changes provide major opportunities and challenges. Increased animal protein supplies provide precious micronutrients and overcome protein malnutrition. More importantly, the demand for more protein supplies provides major opportunities for the owners and producers of animals, intensification of agriculture, and improved livelihoods of the poorest of the poor. However, these challenges are not without problems and include public health concerns, serious environmental hazards arising from, for example, intensive, peri-urban dairy production. These concerns will need to be concurrently addressed with the primary task of producing more foods of animal origin. These issues have raised questions about the contribution by components of the animal industries, the efficiency of individual animal production systems and their capacity to meet the increased demand for animal products. A question of relevance that is being asked is what are the opportunities and challenges for significantly improving current productivity from animals in the future. This paper focuses on the nature of the production resources, and the prevailing animal production systems and trends. It discusses potential opportunities to increase animal protein supplies, and emphasises the need for increased research funds and institutional commitment to support productivity enhancement.

Role and contribution of animals and trends

The individual animal populations in Asia are characterised by diversity and considerable size. These are distributed widely across the preponderance of small farms, which are the reservoir of a large proportion of the main animal species (buffaloes, cattle, goats, sheep, chickens, pigs and ducks). These together form an important economic and ecological niche throughout Asia. Their functions and contribution are numerous. They are consistently and widely owned by small farmers for a variety of advantageous reasons (Devendra, 1983; Chantalakhana, 1990):-

- Diversification in the use of production resources and reduction of socio-economic risks
- Promotion of linkages between system components (land, crops and water)
- Generation of value-added products (e.g. meat, milk, eggs and skins)
- Income generation, investment, insurance and economic security
- Supply of draught power for crop cultivation, transportation and haulage operations
- Contribution to soil fertility through nutrient cycling (dung and urine)
- Contribution to sustainable agriculture, and environmental protection
- Prestige, social and recreational values, and
- Development of stable farm households.

Animal production systems

Animal production involves both non-ruminants and ruminants and a variety of systems integrated with crops. Systems vary as a function of agro-ecological zone and the intensity of farming operations. Development of these systems has considerable potential, the benefits being associated with the complementary interactions of the subsystems, the products from which are additive. Table 14.1 gives an idea of the diversity of available animals and their wide distribution across ecosystems and by sub-region.

Animal production trends in Asia are of two categories (Steinfeld, 1999):-

Table 14.1. Distribution of domestic animals by ecosystem and sub-region in Asia (Devendra, 1995a).

Sub-region	Agroecosystem and animal species										
	Lowland irrigated			Lowland/upland rainfed			Semi-arid and arid			Highland	
	Buffalo/ cattle	Goats/ sheep	Pigs/ poultry/ ducks	Buffalo/ cattle	Goats/ sheep	Pigs/ poultry/ ducks	Buffalo/ cattle	Goats/ sheep	Pigs/ poultry/ ducks	Buffalo/ cattle	Goats / sheep
China	***	*	***	**	***	***	*	***	*	**	***
Hindu-Kush	***	*	**	**	***	*	*	-	-	*	***
South Asia	***	*	**	**	***	**	*	***	-	*	*
Mekong countries	***	*	***	**	**	***	**	*	-	*	*
South East Asia	***	*	***	**	**	***	*	**	*	*	*

(a) Modern, demand-driven and capital intensive non-ruminant sector which produces poultry meat, eggs and pork. Some milk is also produced from dairy cattle and buffaloes. These are mainly industrial and peri-urban systems, which are very efficient and have good market access. Increasing intensification and concentration of animals increases pollution and disease risks to humans.

(b) Traditional, resource-driven and labour intensive ruminant sector, which produces a multitude of services to subsistence farms. Low technology uptake, insufficient market facilities and infrastructure, and small economies of scale are common.

Presently, the non-ruminant pig and poultry industries make the most significant contribution to meet the supply in peri-urban production systems. The application of major advances in non-ruminant nutrition enables maximum *per animal* performance as well as rapid expansion of the enterprise in which the efficiency of feed conversion is high. With ruminants by comparison, overall meat production continues to come mainly from the slaughter of numbers rather than improved animals having good growth rates optimum slaughter weights, and short duration to slaughter. Without exception, this is the trend with beef from cattle, buffalo meat, and meat from goats and sheep. In addition to this, about 150,000 million t of meat from male buffaloes that are past their productive worth is produced in India and exported to countries in South East Asia and the Middle East.

Animal production systems and trends in Asia are influenced to a very large extent by a strong demand-led process. The problem is essentially one of inadequate supplies in which the demand is dictated by a rapidly changing external environment, which is associated with the following factors:

- Population growth
- Urbanisation
- Income growth
- Efficiencies in natural resource management (NRM)
- Supplies being unable to meet the demand, and
- Changing consumer preferences

Increased human population growth together with increasing urbanisation, will significantly drive the demand for animal foods. The increases are awesome, and at projected human population growth rates of 0.7%, 1.6% and 1.4% per year up to year 2010, in China, India and Asia, the population increases by 2010 will be 33%, 18% and 12% respectively. It is significant to also note that of these, between 47-57% of the population will be economically active in agriculture (Alexandratos, 1995). Among South Asian countries, India (60%) and Nepal, (88%), had the highest percentage of the human population active in agriculture.

Table 14.2 presents the trends in *per capita* meat and milk consumption. With the exception of South Asian countries, meat consumption significantly increased, especially in East Asia. Milk consumption has also been increasing. This trend is also consistent with the fact that consumers have been obtaining an increasingly greater share of calories and protein from animal food products in 1993 than before (FAO, 1997). Between 1982-1994, the annual growth rate for total meat consumption was generally high for South East, and East Asia (5.6 to 5.8%) and especially high for China (8.6%). The demand for, and increased consumption of animal foods, is directly related to increased affluence and increased disposable income. At higher levels of income *per capita* consumption of meat levels off because of saturation (Delgado *et al.*, 1999). China and India fall out of this trend because of the very high consumption of pork in the former and religious preferences against meat in the latter. The same authors have also reported that either pork or mutton is the main substitute for beef and the preference increased with increasing income. The situation is serious in that the projected consumption levels in year 2020 are way in excess of production or supply levels. With meat, the shortfall in projected consumption levels in China and other South East Asian countries are about 50-115%. With milk, the supply deficits in China, India and South East Asia are approximately 100%, 89% and 433% respectively.

Table 14.2. Current production and projected consumption trends (1993-2020). (Adapted from Delgado *et al.*, 1999; 2001).

Region	*Per capita production in 1993 (kg)*		*Per capita consumption in 2020 (kg)*		*Annual growth of production 1993-2020 (%)*		*Shortfalls allowing for annual growth with meat*	
	Meat	*Milk*	*Meat*	*Milk*	*Meat*	*Milk*	*Prod.*	*%*
China	33	6	71	16	3.0	5	38	115.0
Other East Asia	24	30	54	29	3.2	1.7	30	125.0
India	5	66	7	104	3.5	3.2	2.0	40.0
Other South Asia	8	62	12	78	3.3	3.0	4.0	50.0
South East Asia	16	3	29	12	3.3	2.9	13.0	81.3
World	34	93	44	87	2.0	1.5	10.0	29.4

While non-ruminant production systems will continue to contribute to the major share of meat and egg production to meet projected human needs, by comparison, considerable opportunities exist for increasing productivity in the ruminant sector in tandem with the need to increase animal protein supplies, food security, reduce poverty and efficient NRM.

Demand-driven consequences

The projected needs for more foods of animal origin in Asia have a number of demand-driven consequences, which also need to be addressed. These include *inter alia*:

- Stress on use of natural resources
- Emphasis on increased productivity *per animal*
- Improved efficiency in feed resource use
- Intensification of animal production systems
- Increased concentration of animals in peri-urban areas
- Increased disease risks, pollution and human health issues, and
- Urbanisation associated with increased consumption of pork, mutton and poultry.

The efficient use of natural resources (land, water, crops and animals) will need to ensure productivity enhancement and environmental integrity (Figure 14.1). Among the natural resources, land use systems will be paramount. In the past, emphasis and the major productivity of cereals through the "Green Revolution" came mainly from the irrigated arable areas. With it came many benefits and prosperity, especially to rice farmers, but the higher agricultural growth exacerbated poverty and food insecurity among the poor in rainfed areas. Increased concentration of animals in peri-urban areas, and intensification of animal production systems refer mainly to non-ruminants (pig and poultry production) and to a less extent to dairy production. The former are mainly vertically integrated systems based on imported animals, feeds, technology and machinery in which the use of locally available raw materials is minimal.

Mixed farming systems

Mixed farming systems, synonymous with crop-animal systems, are varied and involve integrated production resources. These systems form the backbone of Asian agriculture and are one of three primary production systems identified in a description of world livestock systems (Sere and Steinfeld, 1996). Both ruminants and non-ruminants are involved, and the choice of one or more species is dependent on overriding influence of preference, market dictates, potential to generate income, contribution to crop cultivation and contribution to livelihoods.

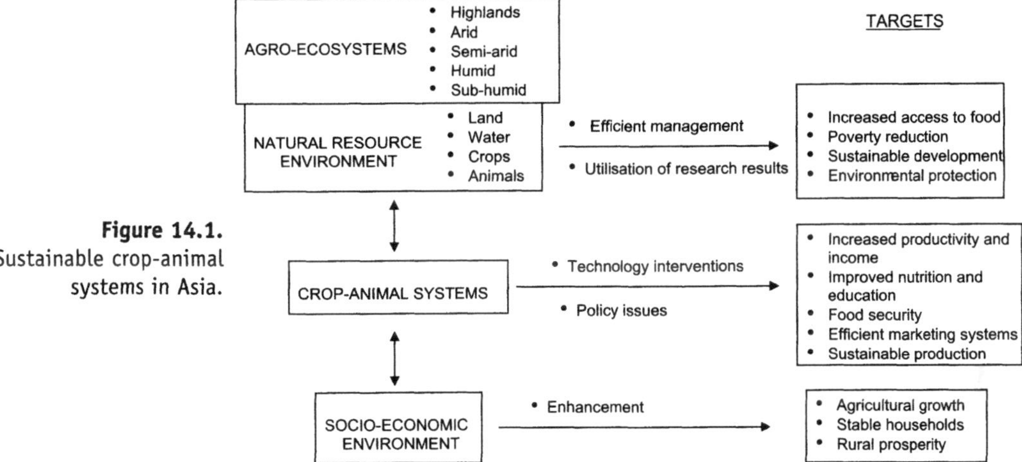

Figure 14.1. Sustainable crop-animal systems in Asia.

The integration of various crops and animals enable synergistic interactions, which have a greater total contribution than the sum of their individual effects (Edwards *et al.*, 1988). Additionally, both ecological and economic sustainability are addressed in a mutually reinforcing manner. Such integrated systems are especially well developed in East and South East Asia. An overview of their potential importance and relevance to small farms in Asia, and description of the distinctive characteristics has been reported (Devendra, 1995a; Devendra, 1996). These include *inter alia:*

- Diversified and integrated use of the production resources, mainly crops and animals.
- Use of both ruminants (buffaloes, cattle, goats and sheep) and non-ruminants (chickens, ducks and pigs).
- Animals and crops play multi-purpose roles.
- Crop-animal-soil interactions are varied and have socio-economic and ecological implications.
- Low inputs use, indigenous and traditional systems
- Involves the three main AEZs (highlands, semi-arid and arid tropics, and sub-humid/humid tropics).

In Asia, it has been reported that mixed farming provided 90% of the milk, 77% of the ruminant meat, 47% of pork and poultry meat, and 31% of the eggs (Steinfeld, 1998). Past growth trends suggest that mixed farming systems grew half as fast (2.2% per year) compared to industrial systems (4.3% per year), and three times as fast as that of pastoral systems (0.7% per year). The data suggest that ruminant production in mixed farming systems will continue to be an important one, within which considerable opportunities exist for improving the efficiency of feed resource use, production systems, and animal performance.

These trends in production outputs are expected to continue in Asia, emphasising the need for improved management of the natural resources. Although industrial systems are expected to grow at a faster rate, good opportunities exist for expanding crop-animal systems, especially into the rainfed lowland and upland areas where research and development has been generally sparse. This is especially significant in the context of the availability of approximately 141 x 10^6 ha of land or 43% of the total arable land, and the presence of 51% of the total human population, and 51-55% of cattle and small ruminant populations in Asia (TAC, 1992).

Animal production within the mixed farming systems is dependent to a very large extent on the efficient use of the available feeds. With ruminant production systems, there are three categories: extensive systems; tethering systems combining arable cropping (roadside, communal and arable grazing systems, tethering, and cut-and-carry feeding); and systems integrated with tree cropping. These production systems are unlikely to change in the foreseeable future (Mahadevan and Devendra, 1986; Devendra, 1989). New proposed systems and returns from them would have to be demonstrably superior and supported by massive capital and other resources. However, as is evident in dairying that there will be increasing intensification and a shift within systems, especially from extensive to systems combining arable cropping, induced by population growth and the fact that population density and intensity of land use are positively correlated (Boserup, 1981). This situation is increasingly likely with decreasing availability of arable land in many parts of South East Asia. The principal aim should therefore be improved feeding and nutrition, in which the objective is maximum use of the available feed resources, notably crop residues and low quality roughages, and also various leguminous forages as supplements.

Rainfed environments

Arable land is limited especially since existing irrigated lands are overused, the potentially important rainfed lands will become increasingly important for productivity enhancement in the future. The rationale and justification for targeting the rainfed areas in Asia is related to the several reasons to meet the need to increase productivity from animals to meet the projected human needs. Productivity growth is stagnant since the biggest gains have already been made, and the potential of rainfed areas for agricultural growth. Pingali and Rosegrant (1994) have reported that the long-term consequences of declines in yield growth in irrigated areas are due to build up of salinity and water logging, micronutrient deficiencies and increased incidence of soil toxicities, soil compaction, decline in soil nitrogen, increased pest build-up and pest-related diseases. Inadequacy of arable land including decreasing availability *per capita* is associated with the following reasons:-

- Demand for agricultural land to meet human needs e.g. housing, recreation and industrialisation
- Expansion of crop production to ceiling levels
- Increasing and very high animal densities
- Increased urbanisation and use of arable land, and
- Growing environmental concerns due to very intensive crop production e.g. acidification and salinisation with rice cultivation, and human health risks due to expanding peri-urban poultry and pig production.

These circumstances thus force a need to look beyond the use of arable lands in mainly irrigated areas, and to focus much needed attention on the more difficult rainfed areas. Consideration must therefore be given to improve use of these areas and the management of natural resources therein. Interdisciplinary research involving integrated NRM in these environments will be especially challenging, but the rewards will be significant. Additionally, it will be equally challenging to ensure that there will not be any environmental problems such as those that have occurred in the irrigated areas.

Given the fact that these areas are overused, attention now needs to shift to the rainfed areas. TAC (1992) has reported that the available rainfed area is about 82% of the land area in Asia in the priority agro-ecological zones (AEZs). The largest areas are found in the arid/semi-arid zones and also humid/sub-humid zones, which are also the priority AEZs. Within these, two broad areas are recognised: rainfed lowlands and rainfed uplands. The two areas are a continuum, with the former having greater opportunities for crop cultivation because of increased soil moisture and less fragility. The characteristics of the lowlands and uplands have been reported (Devendra *et al.*, 1997). Currently, data for the AEZs in South East Asia are not available, so the statistics presented include South Asia. Most of the humid and sub-humid lowlands are found in South East Asia, while the lowlands of South Asia are semi-arid or arid. The lowlands have larger areas of arable and permanent cropland, which account for the greater crop production in these areas. Unfortunately, no data were available for the sizes of buffalo and non-ruminant (pig and poultry) populations, but relatively large numbers of more than 75% of buffaloes and about 70% of non-ruminants are found on small farms.

Poverty and the rural poor

The rain-fed environments are also characterised by abject poverty and the home of the poorest of the poor. Poverty continues to be persistent although absolute numbers have fallen. Within Asia, the largest numbers of poor people are found in India, despite poverty having been reduced from about 55 % in the 1970's to about 35 % today. It is estimated that about 678 million of the rural poor keep livestock in developing countries. These represent two thirds of the rural poor. With reference to the relative numbers of resource-poor by production systems, Livestock in Development (1999) have reported that of the three types of systems, the mixed crop-animal systems have the largest numbers of the rural poor (57 %), followed by the landless (23 %), and the extensive system (23 %). The role and contribution of animals to the resource poor, and the links to and poverty and food insecurity, has recently been reviewed in detail in the context of opportunities for improved livelihoods through efficient NRM (Devendra and Chantalakhana, 2002).

Crop-animal interactions

Crop-animal interactions benefit small farmers and contribute to the sustainability of mixed farming systems (Devendra and Thomas, 2002). These interactions are varied, but are very important from the standpoint of farm productivity and sustainability of mixed farm systems. Draught animals speed up operations such as ploughing and cultivating, and increase the land area prepared for cropping. The vast majority of farmers in the region do not have the resources to replace draught animal power with tractors. The environmental benefits and economic savings to Asian nations through the use of draught animal power has been highlighted by Ramaswamy (1985), who estimated that it would take 30 million tractors to replace some 300 million draught animals on small farms. The use of renewable animal power instead of non-renewable fossil fuels and tractors has, amongst other things, reduced carbon dioxide and carbon monoxide emissions into the atmosphere.

Economic benefits

The economic benefits of improved NRM and impacts in terms of case studies and the application of potentially important technologies have recently been reviewed (Devendra, 1996). The benefits are the result of crop-animal interactions of draught power, manure, feeds and weed control that enhance productivity and income generation of small farmer as well as contribute to the sustainability of mixed farming systems (Devendra and Thomas, 2002). It is not intended therefore to review the data again other than to summarize from these studies the extent of the benefits from animals within a variety of crop animal systems. Table 14.3 summarises the situation from 11 case studies. The data in Table 14.3 have been adapted to reflect the benefits due to interventions in each system to emphasize the value of integrating animals into the systems.

Opportunities and challenges for research in crop-animal systems

The decreasing availability of arable land, as human population densities increase, is a cause for much concern in the debate on the future of crop-animal systems. Arable land is projected to decrease from between 0.17 and 1.0 ha/person in 1988/90 to 0.05 and 0.30 ha/person by 2010 (FAO, 1989). Yet Alexandratos (1995) has calculated that rainfed land suitable for cropping in Asia, and presently not utilised, approximates to 118 million ha. Not only would crop production be increased, but also the potential supply of crop residues and agro-industrial by-products for animal feed. In South East Asia for example, the main focus has been on crop production in the irrigated rice areas, which are already intensively used and over-populated with people. The irrigated mixed farming systems in the humid areas have shown the greatest increases in productivity. This is the result of the strong growth in demand for animal products and better access to feed resources and other types of infrastructure. Devendra *et al.* (1997) have argued that, to further increase crop production in the future, emphasis needs to be given to the neglected rainfed lowland and upland agro-ecological zones. Land is available and advantage can be taken of the significant populations of animals on the mixed farms to integrate them into nutrient and energy cycles that enhance the sustainability of the food crop systems. Also, there is greater poverty in these zones than in the irrigated areas and in the uplands particularly, the ecosystem is fragile with much natural resources degradation. To address these issues, the International Livestock Research Institute (ILRI) is involved with the research project which is aimed at increasing the contribution of livestock to improve productivity in rainfed areas in 5 countries (China, Indonesia, Philippines, Thailand and Vietnam) in South East Asia in a network project called Crop-animal Systems Research Network (CASREN). Good progress has been made to characterize the environment, identify the major constraints which can be overcome by the introduction of appropriate interventions which can increase farm productivity and income, as well as promote the development of sustainable crop-animal systems (Devendra and Pezo, 2002).

The integration of crops and animals places less dependence on the natural resources base than if they are produced separately. Aquaculture has the potential to improve the sustainability and income

Table 14.3. Summary of examples of extent of benefits from animals in improved crop-animal systems in Asia.

Type of crop-animal system	*Country*	*Estimated profitability/net income (US $)*		*Source*
1. Improved beef cattle production systems	Malaysia	Free grazing	0.21 cts/day	Sukri and Dahlan (1984)
		Semi-feed lot	0.28 cts/day	
		Benefits	33.3%	
2. Three-strata forage system	Indonesia	Without project:	106	Nitis *et al.* (1990)
		With project:	186	
		Benefit:	75.5%	
3. Oil palm – cattle – goat	Malaysia	With integration:	110.8/ha/yr	Devendra (1991)
4. Improved grass-legume pastures in coconut plantations integrated with cattle	Philippines	Net profit	US$ 510	Deocareza and Diesta (1993)
5. Small ruminants under coconut, (1991-1994)	Philippines	Net income with project:		PCARRD (1994)
		Sheep:	25	
		Goats:	35	
		Net income with project:		
		Sheep:	127	
		Goats:	229	
6. Food crops, rubber and animal production system (1 cow, 3 goats and 11 chickens)	Indonesia	Net farm income		CRIFC (1995)
		Without project:	24-81	
		With project:	124-138	
		Benefit:	149.5%	

Table 14.3. Contd.

Type of crop-animal system	Country	Estimated profitability/net income (US $)		Source
7. Rainfed rice, upland crops and cattle fattening (1984-1992)	Philippines	Average net income/farm		Sevilla *et al.* (1995)
		Before project:	733	
		After project:	1130	
		Benefit:	54.2%	
8. Baby corn-beef cattle integrated production	Thailand	Benefit	797.3/steer	Prucsasri and Thanomwongwathana (1995)
9. Rice – pigs – fish – duck – vegetable system	Vietnam	Without project:	661	Thien *et al.* (1996)
		With project:	2479	
		Benefit:	275%	
10. Integration of leguminous hedgerows on steep slopes (sloping agriculture land technology)	Philippines	Net profits	865.8-1940.1	Laquihon *et al.* (1997)
11. Rice + fish + duck	Indonesia	Rice alone:	949.8	Suriapernama *et al.* (1998)
		With duck + fish:	2059.7	
		Benefit:	116.9%	

generation of small farms, when it is fully integrated with other enterprises and household activities so as to allow farm families and communities to manage their natural resources effectively. Fish convert crop, livestock and household wastes into high quality protein and nutrient-rich pond mud that can replace fertiliser completely in small vegetable gardens. Aquaculture systems are especially advanced in China (Congyi and Yixian, 1995) and Vietnam (Thien *et al.*, 1998) in terms of practice, efficiency and complementary management of the natural resources. On the other hand, despite some individual success stories, intensive aquaculture has done little to reduce poverty and malnutrition (Prein *et al.*, 1995). Integrated farming systems that include semi-intensive aquaculture can be less risky because, when managed efficiently, they can benefit from synergism among enterprises, diversity in produce and environmental soundness.

Relatively more attention will continue to be given to those mixed farming systems that involve annual crops, because of the importance of rice and wheat as food staples. However, not enough has been done to integrate annual legumes into cereal cropping to develop the potential importance of food-feed systems in Asia (Devendra, Sevilla and Pezo, 2001). The decreased availability of arable land in many areas and the need for more food from animals could encourage further integration of ruminants with tree crops in the upland areas. Some 210 million hectares of land in South East Asia under tree crops are potentially available for this purpose. Associations of tree crops and animals are established farm practice in many developing countries. The expansion and intensification of these systems is a realistic objective, given the extent of farmer experience, the periodic collapse of world prices for plantation commodities and the projected demands for animal products in the future. New technologies to intensify production and better scientific guidelines for managing the components of silvo-pastoral systems are now available that can lead to higher farm incomes and a more protected environment. Future development of these integrated systems will require policy support to encourage the introduction of ruminants to increase their productivity.

Crop-animal systems will see continued intensification and important growth in the future, and that smallholder mixed farms will remain predominant in Asia for some time to come (Devendra, 2002). Important productivity gains can be achieved by further enhancing nutrient and energy flows between the crop and livestock components. Animals, in addition to production, will continue to enhance the natural resources base. The environmental and economic stability of the mixed farming system will make it the prime focus for continuing transfer of technology and development. Important interactions between cropping and livestock production are currently under-exploited, but provide exciting possibilities for technological change in the future. However, the development of sustainable and productive crop-animal systems in the future will require an increased commitment to interdisciplinary research with a farming systems perspective that can focus on whole-farm situations and priority agro-ecological zones. The recent Asian economic crisis clearly demonstrated that mixed farming systems are much more resilient than was imagined formerly. Farmers were able to survive mainly because of their use of local production resources and diversification. This experience has taught them to be more self-reliant and better able to face similar difficulties in the future. The notion of increased population density in areas such as Java is also unlikely to arise, if policies are implemented to redistribute the population. The Government of Indonesia has in place a transmigration scheme, where families are relocated to other islands such as Sumatra, and provided with land in mixed farming initiatives.

Need for increased research investments

The objectives of increasing productivity from animals through the application and wide adoption of appropriate technologies in target agro-ecological zones, as well as the eradication of poverty, will require increased research investments. Past investments in agricultural research have had major payoffs and clearly justify their increased support (Alston *et al.*, 1998), reinvigoration and priority in developing countries (Pardey and Beintema, 2001). Hazel and Haddad (2001) have recently analysed the links between agricultural growth and poverty alleviation and emphasised the need for more funds from agricultural research. These much needed investments, backed by systems perspectives and community-based research and development activities (Devendra, 2000b) can enhance demonstrable impact on improved productivity in rainfed and less-favoured environments. The approaches together can significantly benefit the owners and producers of animals, projected needs of consumers, improve livelihoods of the poorest of the poor, stimulate agricultural growth and environmental sustainability.

Conclusions

Meeting the projected increased human demand for animal proteins in the future is an awesome task. The challenges for pushing productivity growth and increase the supplies requires concerted strategies that involve a high efficiency in NRM that can target specific agro-ecological zones. Given the need to focus on the more complex rainfed environments with their many problems, concerted interdisciplinary research is essential which involve holistic systems, and more importantly, participatory community- based research and development programmes that can resolve the constraints and improve the livelihoods of the poor. These issues necessitate increased research investments that can enhance productivity, efficiency and poverty.

References

Alexandratos, N. (1995). *World agriculture: Toward 2000.* Food and Agriculture Organisation (FAO), Rome, Italy, and John Wiley and Sons, Chichester, UK, 488 pp.

Alston, J.M., Pardey, P.G. and Rosebroom, G. (1998). Financing agricultural research: international investment patterns and policy interventions. *World Development* 26:1975-1071.

Boserup, E. (1981). *Population and technological change: A study of long term trends.* Chicago University Press, Chicago, USA.

Chantalakhana, C. (1990). Small farm animal production and sustainable agriculture. *Proceedings Asian-Australasian Animal Science Congress,* Taipei, Taiwan, 2:39-58.

Congyi, Y. and Yixian, G. (1995). Rice-based crop-animal systems in China. In: *Crop-animal interactions.*Edited by Devendra, C. and Sevilla, C. International Rice Research Institute (IRRI), Los Baños, Philippines, pp. 381-394.

CRIFC (1995). *Final report on crop-animal systems research.* Central Research Institute for Food Crops (CRIFC), Bogor, Indonesia, 61 pp.

Delgado, C., Rosegrant, M.W. and S. Meijer, S. (2001). Livestock to 2020. The revolution continues. *Annual Meeting International Trade Consortium,* Auckland, New Zealand, 38 pp.

Delgado, C., Rosegrant, M., Steinfeld, H., Ehui, S. and Courbois, C. (1999). *Livestock to 2020. The next food revolution.* International Food Policy Research Institute (IFPRI), Washington DC, USA, 72 pp.

Deocareza, A.G. and Diesta, H.E. (1993). Animal production under improved pasture under coconuts. *Proceedings Regional Working Group on Forages,* Khon Kaen, Thailand, pp.183-193.

Devendra, C. (1983). Small farm systems combining crops and animals. *Proceedings 111 World Conference on animal production,* Tokyo, Japan, Volume 1, pp. 173-191.

Devendra, C. (1989). Ruminant production systems in the developing countries: resource utilisation. In: *Feeding strategies for improved productivity of ruminant livestock in developing countries,* International Atomic Energy Agency (IAEA), Vienna, Austria, pp. 5-30.

Devendra, C. (1991). The potential for integration of small ruminants and tree cropping in South East Asia. *World Animal Review* 66:13-22.

Devendra, C. (1993). Sustainable animal production in small farm systems in South East Asia. *FAO Animal Production and Health Paper No.* 100, 163 pp.

Devendra, C. (1995a). Environmental characterisation of crop-animal systems in rainfed upland and lowland areas. In: *Crop-animal interactions*. Edited by Devendra, C. and Sevilla, C. International Rice Research Institute (IRRI) Discussion Paper Series No. 6, pp.243-263.

Devendra, C. (1995b). Mixed farming and intensification of animal production in Asia. In: *Proceedings of the ILRI/FAO Round Table on Livestock Development in Low-income Countries*, International Livestock Research Institute (ILRI), Addis Ababa, Ethiopia, pp.133-144.

Devendra, C. (1996). Overview of integrated, animals-crops-fish production systems: achievements and future potential. *Proceedings of Symposium on integrated systems of animal production in the Asian Region,* Chiba, Japan, pp. 9-22.

Devendra, C. (1999a). Small ruminant production systems in semi-arid and arid environments of Asia. *Annals of Arid Zone*, Jodhpur, India, 37: 215-232.

Devendra, C. (1999b) Goats: Challenges for increased productivity and improved livelihoods. *Outlook on Agriculture* 29 : 215-226.

Devendra, C. (2002). Crop-animal systems in Asia: future perspectives. *Agricultural Systems* 71:179-186.

Devendra, C. and Chantalakhana, C. (1996). Development of sustainable crop-animal systems in Asia. *Proceedings 6th Asian-Australasian Animal Science Congress*. Edited by Bunyawejchewin, P., Sangdid, S. and Hangsarat, K. Animal Husbandry Association of Thailand, Volume 1:21-39.

Devendra, C and Chantalakhana, C. (2002). Animals, poor people and food insecurity: opportunities for improved livelihoods through efficient natural resource management. *Outlook on Agriculture* (in press).

Devendra, C. and Pezo, D. (2002). Improvement of crop-animal systems and rainfed agriculture to food security and livelihoods in South East Asia. *Proceedings of international symposium on sustaining food security and managing natural resources in Southeast Asia – Challenges for the 21st Century.* Eiselen Foundation, Ulm, Germany, pp. 129-131.

Devendra, C., Sevilla, C. and Pezo, D. (2001). Food -feed systems in Asia. *Asian-Australasian Journal of Animal Sciences* 14: 733-745.

Devendra, C., Thomas, D., Jabbar, M. A. and Kudo, H. (1997). *Improvement of livestock production in crop-animal systems in the rainfed agro-ecological zones of South East Asia*. International Livestock Research Institute (ILRI), Nairobi, Kenya, 107 pp.

Devendra, C., Thomas, D., Jabbar, M.A. and Zerbini, E. (2000). *Improvement of livestock production in crop-animal systems in agro-ecological zones of South Asia*. International Livestock Research Institute (ILRI), Nairobi, Kenya, 108 pp.

Devendra, C. and Thomas D. (2002). Crop-animal interactions in mixed farming systems in Asia. *Agricultural Systems* 71: 27-40.

Edwards, P, Pullin, R.S.V. and Gartner, J.A. (1998). Research and education for the development of crop-livestock-fish farming systems in the tropics. *ICLARM studies and Reviews* No. 16, 53 pp.

FAO (1989). Sustainable agricultural production. Implications of international agricultural research. *FAO Research and Technology Paper 4*, Food and Agriculture Organisatin (FAO), Rome, Italy, xi + 31 pp.

Gryseels, G., Groenewold, J.P. and Kassam, J. (1997). *TAC Database for quantitative analysis of CGIAR priorities and strategies*. TAC Secretariat, Rome, Italy.

Hazell , P and Haddad, L. (2001). Agricultural research and poverty reduction. *Food, Agriculture and Environment Discussion Paper* No 34, Washington D C, International Food Policy Research Institute (IFPRI),Washington DC, USA, 41 pp.

Laquihon, G.A., G. Suico and Laquihon, W.A. (1997). Integration of salt management of crop livestock in slopeland areas: the case of "super" SALT (sloping agricultural land technology): *Proceedings international workshop on sustainable crop-livestock integration in sloping lands of Asia*. Davao, Philippines, 21pp.

Livestock in Development (LID). (1999). *Livestock in poverty-focused development*. Livestock in Development, Crewkerne, UK, 95 pp.

Mahadevan, P. and Devendra, C. (1986). Present and projected ruminant production systems of South East Asia and the South Pacific. In: *Forages in South East Asia and the Pacific*. ACIAR Proceedings No. 12, pp. 1-6.

Nitis, I.M., Lana, K., Sukanten, W., Suarna, M. and Putra, S. (1990). The concept and development of the three strata forage system. In: *Shrubs and tree fodders for farm animals*. Edited by Devendra, C. International Development Research Centre (IDRC), Ottawa, Canada, pp. 92-102.

PCARRD (1994).*Recommendations for sustainable integrated small ruminants-coconuts systems*. Series No. 77, Philippine Council for Agriculture Research and Resources Development (PCARRD), Los Baños, Philippines, 57 pp.

Pardey, P.G. and Beintema, N.M. (2001). *Slow magic. Agricultural R & D a century after Mendel*. International Food Policy Research Institute (IFPRI), Washington DC, USA, 30 pp.

Pingali, P.L.and Rosegrant, M. 1994. *Confronting the environmental consequences of the Green Revolution in Asia*. Environment and Production Technology Division Discussion Paper No. 2, International Food Policy Research Institute (IFPRI), Washington DC, USA.

Prein, M., Lightfoot, C. and Pullin, R.S.V. (1995). ICLARM's approach to the integration of aquaculture into sustainable farming systems. *Global agenda for livestock research*. Edited by Devendra, C. and Gardiner, P. Proceedings of the Consultation for the South-East Asia Region at IRRI, Los Banos, The Philippines, 10-13 May 1995. International Livestock Research (ILRI), Nairobi, Kenya, pp. 117-125.

Prucsasri, P. and Thanomwongwathana, P. (1995). Integration of beef cattle and baby maize. In: *Proceedings Crop-animal interactions*, Edited by Devendra, C. and Sevilla, C. IRRI Discussion Paper Series No. 6, International Rice Research Institute (IRRI), pp. 241-248.

Ramaswamy, N.S. (1985). Draught animal power-socioeconomic factors. In: *Draught animal power production*. Edited by Copland, J.W. ACIAR Proceedings No.10, ACIAR, Canberra, Australia, pp. 20-25.

Sere, C. and Steinfeld, H. (1996). World livestock production systems: current status, issues and trends. *Animal Production and Health Paper* No. 127, Food and Agriculture Organisation (FAO), Rome, Italy, 82 pp.

Sevilla C.C., Carangal V.R. and Ranola, R.F. (1995). Development of crop-animal research methodology for lowland rainfed area in Pangasinan, the Philippines. In: *Crop-animal interactions*. Edited by Devendra, C. and Sevilla, C. IRRI Discussion Paper Series 6. International Rice Research Institute (IRRI) Manila, Philippines. pp. 265-286.

Steinfeld, H. (1998). Livestock production in the Asia and Pacific region: Current status, issues and trends. *World Animal Review* 90:14-21.

Steinfeld, H. (1999). The industrialisation of livestock production in the light of the Asian economic crisis. *Proceedings of the workshop on the implications of the Asian economic crisis on the livestock industry,* Food and Agriculture Organisation (FAO), Bangkok, Thailand, pp. 27-40.

Sukri, M. I. and Dahlan, I. (1984). Feedlot and semi—feedlot systems for beef cattle fattening among smallholders. *Proceedings 8th annual conference Malaysian Society of Animal Production*, Genting Highlands, Pahang, Malaysia, pp. 74-78.

Suriapernama, S., Syamsiah, T., Fagi, A.M. and Atunadja (1998). Optimase daya dukung lahan dengan ristem minapadi-itek pada lahan dengan ristem minapadi-itek pda saevah beririgasi. Simposium Tananam Pangan 11, Bogor, Indonesia, pp. 21-23.

TAC (Technical Advisory Committee) (1992). *Review of CGIAR of Priorities and Strategies Part 1*, TAC Secretariat, Food and Agriculture Organisation (FAO), Rome, Italy, 250 pp.

Thien, N., Cong Quoc, N., Xuan Tuyen, D. and Sasaki, M. (1996). Rice-fish-duck-pig production in Vietnam. *Proceedings*

Symposium on integrated systems of animal production in the Asian region, Chiba, Japan, pp. 53-76.

15

Livestock and the poor: issues in poverty-focused livestock development

C.Heffernan
Livestock Development Group, School of Agriculture, Policy and Development, The University of Reading, Earley Gate, PO Box 237, Reading RG6 6AR,UK

Introduction

Over the course of the last decade, the fight against global poverty has galvanised the world community. Development projects and programmes are increasingly justified and funded solely on their potential to aid the poor. Equally, the lessons learned from previous drives to eradicate poverty appear to have been taken on board. It has been recognised that to lower poverty levels, it is first imperative to understand the lives and livelihoods of the poor. Therefore, current approaches generally focus upon the multitude of activities that poor households pursue. Animal husbandry is one such activity. Indeed, it has been estimated that at least one third of the poor rear livestock (LID, 1999). Therefore, livestock keepers are one of the largest subsets of the global community of the poor.

In recognition of the importance of livestock to the livelihoods of the poor, donors, governments and NGOs have channelled resources to the livestock sector. Early projects tended to be top-down, technology-driven interventions while more recently the focus has been on holistic, participatory and community-based projects and programmes. Indeed, the role of livestock in poverty alleviation has generally mirrored current conceptual notions regarding the state of poverty itself. For example, during the 1970s, poverty was generally deemed a quantifiable condition, described by first, income and later consumption deficits. Livestock projects, during this period, were often justified by the argument that a strong livestock economy would provide trickle down benefits to the poor. Large-scale projects and programmes aimed at improving productivity, at the national level, were common. In the 1980s, notions of poverty as a deficit of consumption began to receive more attention. Definitions of poverty began to include notions of food and livelihood security. Consequently, support for livestock was included as part of wider packages for community development.

During the 1990s, with the advent of Participatory Poverty Assessments (PPAs), the discourse on poverty subtly changed again with the psychological aspects of being poor increasingly emphasised. Feelings of powerlessness, vulnerability and increased fear and anxiety were frequent findings in participatory poverty assessments (World Bank, 2000). Again, another livestock development trend became apparent and projects were increasingly justified on the grounds that livestock could enhance well-being and decrease a household's vulnerability to shocks and disaster. The inference here was that livestock could ameliorate some of the adverse psychological impacts of poverty.

Most recently it has been recognised that poverty is a 'multi-dimensional' phenomena with differing implications for the individuals and communities involved. Current descriptions of poverty focus on the consequences of deprivation (World Bank, 2000). Thus, for the person involved, poverty is often equated to ill-health and hunger, lack of choices and opportunities, low education levels and high mortality rates in addition to a lack of access to capital. Further, at the community level, the poor are often deprived of services and are faced with political and institutional structures, which at best are not geared to addressing their needs and at worst are biased and discriminatory. Livestock development, again, has reflected the conceptual changes. Programmes, at the national level, are often aimed at strengthening institutional frameworks. Conversely, at the community-level, the majority of interventions attempt to enhance food and livelihood security.

Nevertheless, within the livestock sector the transformation from top-down, technology driven projects to more community-based initiatives has not been entirely seamless. The track record of both past and present livestock development projects is mixed. Indeed, a recent review concluded that the majority of animal health projects were not having their intended impact on the poor (LID, 1999). Reasons offered for the poor performance included the lack of a poverty focus and failure to deliver the outputs (*ibid.*).

More fundamentally, however, the overall lack of impact may be due to the continuing tension between the two, often conflicting, aims of livestock development: technological inputs to increase production vs. community development to reduce poverty. Many practitioners believe the problems of the poor may be considerably aided by technological solutions and that community development approaches cannot be scaled-up. Conversely, for the proponents of community development, many technologies are viewed as lacking relevance to the wider problems of the poor and as such are doomed to failure. Consequently, within the livestock sector, many practitioners belong to either upstream or downstream philosophies and livestock research and development activities broadly follow the division.

Furthermore, at the global level, the livestock sector is undergoing rapid transformation. Indeed, never before in history, has the demand for livestock products been at such a high level. Nor has the sector been exposed to as many negative and positive driving forces. As de Haan *et al.* (2001) notes:

> "The global livestock sector is changing fast. With a strong and growing demand, rapid institutional and macroeconomic policy changes, and a fundamental shift in the functions of livestock, there is a significant danger of the poor being crowded out, the environment eroded, and global food security and safety compromised."

Thus, livestock development is under increasing pressure to address the rapidly changing needs and demands of both the poor and the expanding global population.

Consequently, it is apparent that livestock development is now at a crossroads. Although more and more evidence is available to illustrate the importance of livestock in poverty alleviation, projects and programmes are mired in less than positive outcomes. Given the problems of the poor, there is an urgent need for a new paradigm of livestock development, which incorporates both people-centred and technological solutions to the specific needs of poor livestock keepers.

Therefore, the following paper is divided into three sections. The first section offers a profile of poor livestock keepers and a definition of poor livestock keeping households. In the second section, a new approach to poverty-focused livestock development is detailed. Finally, in the closing section, the wider forces predicted to impact development practice and policy, within the livestock sector, are discussed.

Profiles of the poor livestock keepers

As a target group, poor livestock-keepers represent a dynamic spectrum, varying from households where livestock comprise only a small portion of their livelihood activities to those in which livestock are the main component (Heffernan and Misturelli, 2000). Poor households also have a varying dependency upon livestock off-take and products. Therefore, attempts to categorise poor livestock keepers by either the number of animals owned or the household dependency upon off-take may be misrepresentative. For example, Swift (1988) describes a pastoralist household as one where over 50% of the gross revenue is derived from livestock, or equally, where 15% or more of the total energy consumed is derived from livestock products. However, the definition excludes many destitute herders who rely on alternate

intensive monogastric and poultry production in peri-urban areas. Conversely, for the poor, 'landlessness' is rather a more straightforward concept and denotes a lack of access to land. Landless livestock keepers are present in all livestock production systems in developing countries. Therefore, the above classification system, while describing global livestock production, neither details the types of livestock keepers within each system nor outlines the production systems important to the poor.

Poor livestock keepers live on the margins of livestock production systems. For example, in pastoralist production systems, there are generally three distinct populations of the poor, which vary dramatically in their herd management and husbandry strategies. First, there are poor pastoralists who remain active in the livestock economy (albeit at very low levels) and continue to live a more traditional lifestyle in the rural areas. Conversely, the second population of the poor has settled around towns and settlements. In these cases, livestock are one of a range of activities that are pursued. Finally, the third category of the poor resides in peri-urban settlements but via hired or family labour maintain herds in the rural areas. Equally, there is a difference among the groups in the primary species kept. Poor pastoralists who remain in rural areas generally maintain small herds of cattle whereas for their more urban counterparts production tends to concentrate on smallstock. Consequently, the above system fails to capture the nuances needed to specifically characterize poor livestock keepers. Hence, a new typology of livestock production for the poor is required.

As such, Table 15.2 describes a simple typology based upon three broad types of production systems: pastoralist, subsistence farming and urban. The intention is not to strictly define each of the different characterizations but rather to offer a broad framework which describes the many and varied types of poor producers.

As Table 15.2 illustrates, poor livestock keepers keep a wide variety of species and practise a number of different husbandry methods. Where the poor differ greatly from better off producers, however, is in access to inputs and resources for livestock production. Indeed, studies have demonstrated the direct relationship between land and livestock ownership patterns. For example, de Lasson and Dolberg (1985) demonstrated that changes in the intensity of land use changed livestock ownership and production patterns in Asia. Among their study set in India, the loss of a bullock related to losing more than an acre per household, while gain of bullocks related to an increase in land holdings. The poor may also be differentiated by their vulnerabilities. Each of the different production systems has a variety of factors that while negatively impacting well-off producers may be devastating to the poor. For example, drought for a poor pastoralist

Table 15.2. Typology of poor livestock keepers.

Livestock production system	*Characterisation of the Poor*	*Location of Poor producers*	*Livestock species*	*Herding/ Husbandry patterns*	*Vulnerabilities*
Pastoralist	Lack of access to labour for livestock keeping	Rural	Cattle, smallstock, camelids, yak	Migratory	Drought, terms of trade, political instability, poor access to markets, technologies and innovations.
		Peri-urban	Mainly smallstock	Mainly around settlements	Drought, terms of trade: livestock vs. grain
		Peri-urban	Mainly smallstock	Absentee owners, herding by relatives,hired labour	Lack of access to productive resources of animals, theft
Subsistence farming	Small land sizes, land rental, lack of resources	Rura	Cattle, buffalo, smallstock, poultry	Tether, cut and carry, range	Drought, cost of inputs, access to services, population pressures
		Peri-urban	Dairy, poultry, pigs	Stall fed, cut and carry, roadside	Cost of inputs
Urban	Landless	Urbcn	Poultry smallstock, buffalo, cattle, pigs	Roadside, rubbish foraging, purchased forage	Space for animals, legal framework

will obviously lead to destitution much faster than those with greater herd assets. Equally, adequate space for livestock keeping is a major problem faced by poor producers in urban areas. The legislative environment is also an issue and many of the poor are forced to pay fines and bribes to maintain production.

This diversity in livestock keeping households obviously has an impact on the outcome of livestock development projects and programmes. The following section explores the evidence for pro-poor livestock development.

The impact of livestock development on the poor

There are three potential ways of determining the impact of livestock development projects and programmes on the poor. First, at the global level, information and evidence may be gathered on a project-by-project basis for the nations involved. Second, criteria can be devised to evaluate the overall impact of specific types of projects, e.g. animal health, technology transfer etc. Finally, an individual agency approach may be undertaken with specific institutions offering an assessment of their projects and programmes.

However, a variety of obstacles have been noted with all three approaches. At the global level, co-operation and partnership between the actors and agents involved are often weak and fragmented, hence deriving the impact of specific livestock projects is difficult. More success has been achieved in analysing different forms and types of livestock projects. For example, Oakeley (1999) and Martin (2001) offer overviews of community animal-healthcare projects, and Heffernan *et al.* (2001) performed a large-scale review of restocking projects. Nevertheless, obtaining sufficient information from the actors and agents involved is often difficult with the project-level approach. At the agency level, little information is available in the public domain regarding the impact of specific livestock development projects and programmes. Few critical analyses of livestock sector activities have been performed with the notable exceptions of DFID, the World Bank and IFAD.

Indeed, in 1998, DFID undertook one of the most comprehensive reviews of livestock projects and programmes to date (LID, 1998). Over 800 livestock development projects were reviewed for their impact on the poor. Overall, the authors concluded that the majority of livestock projects and programmes had not had a significant impact on the poor for the following reasons:

1. Technologies were developed but not delivered to the poor.
2. The technologies that were delivered were inappropriate to the poor.

3. In cases where appropriate technologies were successfully delivered, wealthier farmers or herders tended to capture the benefits.

A subsequent report by the same authors, offers the following conclusion (LID, 1999):

> "Our review of project documentation on technical and service-related projects revealed little evidence of widespread sustainable impact on the livelihoods of the poor. Although there are some islands of success, the overall tenor of the literature, donor assessments and evaluation reports that we reviewed is that technical and service projects were not successful at benefiting the poor on a sustainable basis."

The finding was corroborated by de Haan *et al.* (2001) who offered the following in regard to World Bank projects:

> "The livestock portfolio analysis shows that our current World Bank operations still lack a specific policy and environmental focus...This lack of focus is shown by the low level of investment in the poorest regions of the world (central Asia, South Asia and Sub-Saharan Africa) in pastoral development and smallstock, and to some extent, in the low share of investments to improve animal health and nutrition, which are critical constraints faced by the poor."

De Haan *et al.* (2001) further note that since the 1970s there has been a decline in support for livestock projects:

> "Currently, six active agricultural projects are livestock only, and about 50 projects (of a total agricultural portfolio of 270) have livestock components. The decrease in lending is partially in response to the poor performance of the projects during the 1970s and 1980s."

Hence, it is increasingly recognised that livestock projects and programmes have not had their intended impacts. Nevertheless, there is little critical analysis of why these wide-spread failures have occurred. Thus, it is apparent that in order for livestock-based livelihoods to deliver their full potential with regard to poverty alleviation, a new approach to livestock development is required.

Towards a new paradigm of livestock development

In order to advance a new paradigm of livestock development, it is clear that conceptual notions of poverty must first be revisited more explicitly. Recent notions of poverty tend to be descriptive of the many different aspects of being poor. Indeed, current notions of poverty tend to mirror the many facets of the sustainable livelihoods framework. The World Bank (2000) offers the following definition:

> "Poverty is pronounced deprivation of wellbeing...to be poor is to be hungry, to lack shelter and clothing, to be sick and not cared for, to be illiterate and not schooled...Poor people are particularly vulnerable to adverse events outside their control. They are often treated badly by the institutions and states and society and excluded from voice and power in those institutions."

However, it appears that there is an increasing confusion between the consequences of poverty and the state of poverty itself. The distinction is important. For example, a recent World Bank report (World Bank, 2000) disaggregates poverty into areas such as 'health and education', 'income' and 'vulnerability' and 'voicelessness and powerlessness'. However, by focusing on the consequences rather than the condition of poverty, the risk is that interrelationships between causal factors will be lost and with it, a comprehensive understanding of the poor.

Further, the new focus, by attempting to put a more human face on poverty, often portrays the poor as the unwitting victims of a hostile political, institutional, social and economic environment. Societies, cultures, households and communities do not generally define themselves either by their levels of deprivation or by their attendant powerlessness. For the fight against global poverty to succeed, concepts that disregard the capabilities and strengths of the poor may ultimately prove counterproductive. Thus, for poverty eradication strategies to meet the needs of the poor, our viewpoint of poverty must change from one of deprivation to one of supporting the future dreams and aspirations of households, families and individuals involved.

Consequently, it is important, when considering poverty that the capabilities and agency of the poor are not ignored. Inarguably, livestock are a key means to facilitate the potential of the poor. The sale and consumption of animal products decrease the vulnerability of households to normal seasonal food and income deprivations. Therefore, livestock can both fulfil wider food security needs and enhance the nutritional status of the most vulnerable: women, children and the elderly. Keeping livestock also helps to shield households from sudden shocks such as civil war and political instability. Further, animal ownership may increase the ability of households and individuals to participate in social rituals and fulfil social obligations.

Among some societies, livestock ownership is also a form of cultural identity. Livestock are also a key source of collateral for the poor and enable many households to obtain access to capital and business loans. Thus, livestock are an important capital asset, which with careful tending can propel households out of abject poverty and into market economies.

Nevertheless, it is increasingly recognised that in the application of livestock as a means of poverty alleviation, for every benefit, there is a direct cost for the poor households involved. Livestock owners face increased household expenditures for animal healthcare and often fodder and water. Poor marketing infrastructure in many countries also limits the profit margins for farmers and hence disposable income. The need to compete with more developed markets often increases processing costs. In addition, poor households often lack the labour needed for sustainable livestock production (Heffernan and Misturelli, 2000).

Therefore, any attempt to better understand poverty, both within the livestock sector and externally, must address the apparent contradictions and conflicts between the causes and consequences of poverty, the needs and capabilities of the poor and the benefits and costs of livestock production. Consequently, in order to develop effective poverty alleviation mechanisms and strategies, the association between the three elements must be derived. Figure 15.1 describes a proposed relationship between the needs, consequences and increased costs for poor livestock-keeping households.

As Figure 15.1 illustrates, the poor have both wider needs with regard to food security, health and housing and more specific needs related to livestock-related livelihoods such as access to markets, livestock drugs and extension advice. Obviously, first order needs must be secured prior to the livestock-related requirements being addressed. In addition, households must balance subsistence needs against the costs of livestock production such as expenses for feed, medicine, housing, and both the direct and indirect costs of labour. With any of these elements missing, households with livestock are more vulnerable rather than less, to the consequences of poverty, such as increased livestock morbidity and mortality and loss of the productive assets of the herd. Conversely, the cycle may be broken by projects and programmes, which support the capabilities and agency of the poor with regard to livestock, thereby enhancing the benefits of livestock production for the households involved. Nevertheless, livestock production does not occur in isolation. While the poor are largely marginalized from commercial trends, projects and programmes must account for the global factors predicted to impact the livestock sector.

countries, human population expansion is eroding large areas of rangeland, which were traditionally reserved for livestock keeping. As such, production rises will increasingly depend on a greater productivity per animal. Therefore, the extent, to which livestock producers are able to adopt practices to enhance productivity, will become more and more important."

Indeed, a comparison of the productivity achieved under intensive practices with those found among small holders in similar locales, demonstrated that existing technologies can greatly enhance productivity (Delgado *et al.*, 1999). Feeding, animal health, breeding, and management practices all offer potential areas for improvement (Pilling, 2002). While well-known techniques such as improving feed quality can have a marked effect on production levels (FAO, 1999), it is believed that the demands of the 'livestock revolution' will require feed resources beyond the capacity of many mixed farming systems. Hence, farmers will be more reliant on grains, which are likely to be imported (Delgado *et al.*, 1999). Consequently, many of the potential benefits of the expanded market will fall to those producers who are able to adopt production practices considerably different from those found among poor livestock keepers (Pilling, 2002).

Thus, without policies which facilitate the participation of the poor, the danger is that the potential benefits of the 'livestock revolution' will not accrue to the least well-off. It is argued that the removal policies that distort the market in favour of large-scale production would eliminate much of the competitive disadvantage faced by small producers (*ibid.*). Nevertheless, the intensification of animal production, generally places greater demands on livestock services. Research has demonstrated that access to livestock services is a major problem for the poor (Heffernan and Misturelli, 2000; Heffernan *et al.*, 2002). Thus, prior to making conclusions regarding the impact of the 'livestock revolution' on the poor, it is necessary to further examine the opportunities and constraints faced by poor livestock keepers in the face of these global trends.

Conclusions

Livestock keepers are a growing subset of the global community of the poor with unique capabilities and needs. However, poor livestock keepers are not a generic and homogeneous group, in which the application of a single solution is sufficient. Rather, to be effective, projects and programmes must account for the rich diversity of producers at the community, national and international levels. Therefore, a better understanding of the unique needs of the differing groups of poor livestock keepers is required. As such, further research is needed to develop a more detailed typology of poor livestock keepers. To effectively inform policy and practice, such a typology, while documenting basic characteristics, should also include the values, aspirations and attitudes toward livestock keeping.

However, it has been increasingly recognised that while livestock keeping and poverty alleviation are closely linked, livestock development projects have rarely fulfilled their promise with regard to the poor. Therefore, given the lack of positive poverty outcomes, it is clear that a new approach to livestock development is required. To be successful, such an approach must closely account for, and address, the realities faced by the poor. Livestock keeping, unlike other livelihood activities, produces an appreciating asset, but also incurs costs for the households involved. At the household level, understanding the factors that inform decision-making will greatly improve the uptake and impact of projects and programmes, particularly those delivering new technologies.

Finally, it is clear that in the coming decades, poor livestock keepers will be subject to a number of forces predicted to sweep the global livestock sector. Additional research is required to illuminate precisely how these trends will impact the poor and the ways and means that these forces can enhance, rather than further limit, the sustainability of livestock-based livelihoods. Thus, with the appropriate support, livestock can, and should be, a key mechanism in the fight for global poverty alleviation.

Acknowledgements

This publication was an output from the Global Initiative on Livestock Services and the Poor which was informed by the research project R7359 Animal Health Programme funded by the UK Department for International Development for the benefit of developing countries. The views expressed are not necessarily those of DFID.

References

Haan, C. de, Schillhorn van Veen, T., Brandenburg, B., Gauthier, J., Le Gall, F., Mearns, R., Simeon, M. (2001). *Livestock development: Implications for rural poverty, the environment and global food security*. The World Bank, Washington, DC, 96 pp.

Delgado, C., Rosegrant, M., Steinfeld, H., Ehui, S., and Courbois, C. (1999). *Livestock to 2020: The next food revolution*. Food, Agriculture and Environment Discussion Paper 28, International Food Policy Research Institute (IFPRI), Washington, DC, USA, 72 pp.

FAO, (1999). *Poverty alleviation and food security in Asia: Role of livestock*. Food and Agriculture Organisation (FAO), Regional Office for Asia and the Pacific. In: http://www.fao.org.

Heffernan, C. and Sidahmed, A. (1998). *Issues in the delivery of veterinary services to the rural poor.* Paper presented at the Conference on the Delivery of Veterinary Services to the Poor, June, 1998, University of Reading, UK.

Heffernan, C. and Misturelli, F. (2000). *The delivery of veterinary services to the poor: Findings from Kenya*. Report for the Department for International Development (DFID) Animal Health Programme, University of Edinburgh, UK, 81 pp.

Heffernan, C., Misturelli, F. and Nielsen, L. (2001). *Restocking and poverty alleviation: Perceptions and realities of livestock keeping among poor pastoralists in Kenya*. Report for the Department of International Development (DFID), Livestock Production Programme, Natural Resources International Ltd., Aylesford, UK, 87 pp.

Jahnke, H. (1982). *Livestock production systems and livestock development in tropical Africa*. Kieler Wissenschaftsverlag Vauk, Kiel, 253 pp.

Lasson, A. de and Dolberg, F. (1985). The causal effect of landholding on livestock. *Quarterly Journal of International Agriculture* 24 (4):339-354.

Livestock in Development (LID) (1998). *Strategies for improving DFID's impact on poverty reduction: A review of best practice in the livestock sector*. Livestock in Development, Crewkerne,UK, 57 pp.

Livestock in Development (LID) (1999). *Livestock in poverty focused development*. Livestock in Development, Crewekerne, UK, 94 pp.

Martin, M., (2001). *The impact of community animal health services on farmers in low-income countries: A literature review*. VETAID, Edinburgh, UK, 64 pp.

Oakeley, R (1999) Community Livestock Workers: A review of experiences with community-based animal healthcare delivery, *World Animal Review* 93 (2), Food and Agriculture Organisation (FAO), Rome, Italy: www.FAO.org/docrep/x3770t/x3770t06.htm#P0_0

Perry, B.D., McDermott, J.J., Randolph, T.F. Sones, K.R. and Thornton, P.K. (2002). *Investing in animal health research to alleviate poverty*. International Livestock Research Institute (ILRI), Nairobi, Kenya, 140 pp.

Pilling, D. (2002). *Issues in livestock disease prioritisation*. PhD Upgrade document. The University of Reading, Reading, UK, 79 pp.

Seré, C. and Steinfeld, H. (1996). *World livestock production systems: Current status, issues and trends.* FAO Animal Production and

Health Paper 127. Food and Agriculture Organisation (FAO), Rome, Italy, 92 pp.

Swift, J. (1988). *Major issues in pastoral development with special emphasis on African countries.* An informal seminar on socio-economic aspects of pastoral development. Food and Agriculture Organisation (FAO), Rome, Italy, 12 pp.

Wilson, R.T. (1995). *Livestock production systems. The Tropical Agriculturalist.* Series edited by Coste, R and Smith, A.J. Macmillan, London and Oxford, in co-operation with CTA (Technical Centre for Agriculture and Rural Co-operation), Wageningen, The Netherlands, 160 pp.

World Bank, 2000. *World development report 2000/2001: Attacking poverty.* Oxford University Press, New York, 266 pp.

16

GM technology: a tool to benefit livestock production

R.H. Phipps[1] and A. Cockburn[2]
[1]*School of Agriculture, Policy and Development, The University of Reading, PO Box 237, Reading RG6 6AR, UK*
[2]*Monsanto Company, Maris Lane, Cambridge CB2 2LQ, UK*

Introduction

The limitations to livestock production are numerous and varied but are applicable to a greater or lesser extent in both developed and developing countries. Crop factors that limit livestock production include inadequate quality (e.g. crop residues such as maize stover) and quantity of feed resources with an inconsistent supply due to extremes of climate (e.g. low rainfall and high temperatures), the presence of anti-nutritional factors and toxins (e.g. trypsin inhibitors, glucosinolates, gossypol, mycotoxins), and deficiencies of specific nutrients (e.g. amino acids and minerals). These and other factors have been the subject of a number of reviews including those by (Mgheni, *et al.*, 1993; Gardiner and Devendra, 1995). While improved animal genetics have greatly increased productivity per animal (e.g. high genetic merit cows) in developed countries, the low genetic potential of many of the breeds in developing countries is still limiting (Cunningham, 2003). Improved feed resources are usually needed before increased animal productivity can be realized through enhanced genetics. Furthermore the occurrence of animal diseases is a serious limitation in livestock production. This paper will discuss how GM (genetically modified) technology can contribute a way forward with the aim of combining higher yields, improved food and feed quality and increased competitiveness with environmentally and agriculturally sustainable practices.

The "Green Revolution" which combined effects of improved varieties, increased use of fertilizer, irrigation and pesticides was instrumental in allowing world food production to double in the last 35 years. However, within the current agricultural production systems, 5% of global fossil fuels are used for fertilizer production, and synthetic pesticides are used extensively (2.5 million t/year) to try to reduce crop losses (Paoletti and Pimentel, 2000). The annual losses for eight of the major world crops has reached a staggering US$ 244 billion, representing 41% of total annual production. Of this figure 15%,

13% and 13% are attributed to insects, diseases and weeds, respectively (Oerke and Dehne, 1997), with post harvest losses estimated to be a further 10%; in short, only half the food grown becomes available for consumption. As we enter the 21st century, world population is set to increase to 9 billion by 2050 (United Nations, 1996), with the rate of increase being higher in less developed than developed countries. At the same time the rate of crop improvement achieved through conventional breeding is slowing as illustrated by the decline in annual growth in cereal yields reported in Food and Agriculture Organisation of the United Nations <http://www.faostat.fao.org>. In addition the area of land available and appropriate for the production of food and feed resources is at best static and at worst decreasing, often due to wind, water and soil erosion (FAO, 1993). In this scenario Delgado *et al.* (2001) have also estimated that the demand for livestock products will increase dramatically in the next 20 years. The increase will be markedly higher in developing countries (3% per year) where much of the increased demand will be generated from an increasing urban population as compared to developed countries (<1% per year). In order to achieve this projected increase in demand for livestock products there will be a need for a very significant increase in both yield and quality of feed resources including forages, cereal grain and oilseeds, which is further compounded by the increased pressure on land use by increasing livestock production. These changes must be achieved in a manner ensuring safety for the public, animals and the environment. A major challenge for global food and feed production is to maintain or improve its productivity and sustainability while decreasing the environmental impact (Tillman, 1999).

Current position on uptake of GM technology

In the year 2002, six million farmers in 16 countries grew 59 million ha of transgenic crops (James, 2002), with further increases predicted for 2003. The increase from 4 million ha in 1996 to 59 million ha in 2002 represents the fastest adoption rate of any new technology ever brought to agriculture with an accumulative total of 235 million ha grown in this period (Figure 16.1). The principal GM crops grown are soyabean (36.5 million ha), maize (12.4 million ha), cotton (6.8 million ha), and canola (3.0 million ha). With few exceptions these crops have been modified for first generation agronomic input traits such as herbicide tolerance and, or, insect protection. These crops are used in livestock production rations for both non-ruminants and ruminants as either energy and, or, protein feed resources. The use of soya in particular has increased since the use of meat and bone meal as a nitrogen source was banned in Europe. They are included: either in the form of whole crop, for example maize silage; as a specific crop component, such as maize grain; or as co-products, such as oilseed meals, or maize stover. In 2002, 75% of the crops were grown in developed agricultural systems, with 25% being grown in developing countries. There is clear evidence that GM technology is scale neutral, as shown by the success of the smallholder schemes in South Africa and India (Ismael *et al.*, 2002). The second generation of GM crops, which are currently in field trials, have been modified for improved nutritional characteristics for both food and feed.

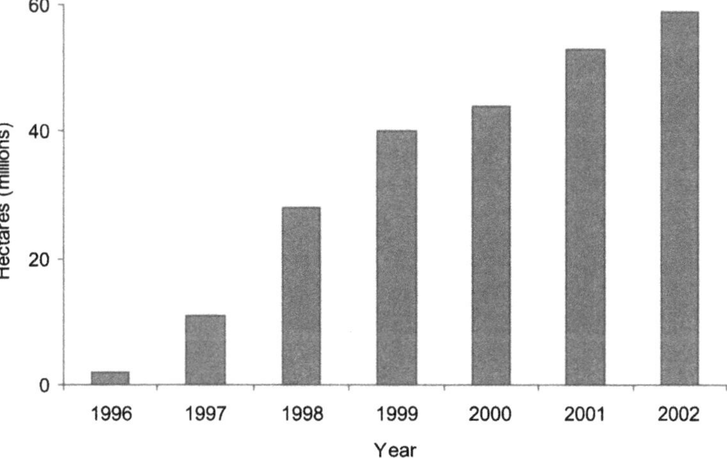

Figure 16. 1. Global area (millions ha) of transgenic crops (Source: James 2002).

Environmental and sustainable agricultural practices

The Bruntland Commission defined sustainable development as "meets the needs of the present without compromising the ability of future generations to meet their own needs", and as such, the long-term success of both ruminant and non-ruminant livestock production must be conducted through environmentally friendly and sustainable agricultural practices. In this context, the reduction in pesticide use has been a high priority for EU policy makers through incentives and management (MAFF, 1999). It is of note that a number of studies have now clearly established that growing agronomic input trait GM crops reduces pesticide inputs in crops that often form the cornerstone of ruminant and monogastric livestock production. Phipps and Park (2002) have recently reviewed this topic and estimated that the introduction of GM varieties of soyabean, oilseed rape, maize and cotton had reduced pesticide use on a global basis by 22 million kg of formulated product in the year 2000. Further studies by Carpenter *et al.* (2002), Gianessi *et al.* (2002), Coyette *et al.* (2002), Kent and Moniem (2002) and May (2003) have all shown that the introduction of GM crops has the potential to reduce significantly the use of pesticides. Perhaps the most comprehensive of these reports is the one produced by Carpenter and colleagues who reported, "a comprehensive review of the scientific literature supports the overall conclusion that the currently commercialized biotech derived soybean, corn and cotton yield environmental benefits".

In addition transgenic crops are sprayed less frequently. In reducing field operations in Canada with transgenic canola, farmers reduced diesel use by 31million litres/annum (Canola Council, 2001) and one of the consequences of this was a significant reduction in emissions of CO_2. It is calculated that the reduction in spraying operations (0.4 operations/ha), the reduction in the need for mechanical weeding and the increase in "No-Till cultivation practice", through growing 45 million ha of GM soyabeans in Argentina since 1996, emissions of CO_2 have been reduced by 1.7 million t. In addition Barnes *et al.* (2000) have estimated that "No-Till" operations will also result in decreased topsoil erosion (37 million t by 2020) and reduce the amount of soil carbon released into the atmosphere where it combines with other gases that contribute to global warming. Soyabean growers in the US have rapidly moved to reduced tillage practices with GM herbicide tolerance cited as the factor which had the greatest impact toward the adoption of reduced tillage or no-till (American Soybean Association, 2001). The reduced input costs and the often-associated increase in yield enhances economic competitiveness. Second generation GM crops will also provide significant environmental benefits. For example crops modified for improved nitrogen digestibility, or those with high phytase or low phytate content, will be a major factor in reducing nitrogen and phosphorous excretion and hence pollution. Products that can manipulate rumen fermentation will also be developed resulting in reduced methane production. The introduction of GM technology which has resulted in a large and significant increase in the use of conservation tillage is associated with reductions in wind and water erosion and further reductions in the use of diesel and, hence, emissions of greenhouse gasses.

Safer feed

In many parts of the world maize is attacked by a variety of insect pests. For example, the European 'Corn Borer' can significantly reduce crop yield and decrease the quantity and quality of food and feed available for both humans and livestock. After the initial insect attack the maize crop is susceptible to secondary fungal infection as the spores enter the wounds left by the insect. Modern biotechnology has produced insect-protected plants that have been genetically enhanced to produce proteins, similar to those produced by the soil bacterium *Bacillus thuringiensis* (*Bt*), which kill the insect pest. Munkvold and Hellmich (1999) first noted a reduction in fumonisin concentration in *Bt* maize grain and subsequent studies in Europe (Pietri and Piva, 2000) have established consistent and significant reduction in total fumonisin (B1, B2 and B3) concentrations in maize grain derived from plants genetically modified for insect protection (Figure 16.2).

While the majority of reports in the literature show compositional and nutritional equivalence when comparing GM and conventional feed ingredients (Clark and Ipharraguerre, 2000) work conducted by Piva *et al.* (2001a,b) noted increased animal performance in both pigs and poultry when diets contained *Bt* maize grain rather than conventional grain (Table 16.1). The increased performance was attributed to a significant reduction in mycotoxin contamination resulting from the use of *Bt* maize grain. Thus, the use of the first generation insect-protected maize varieties can result in not only more grain but safer grain for both livestock and humans

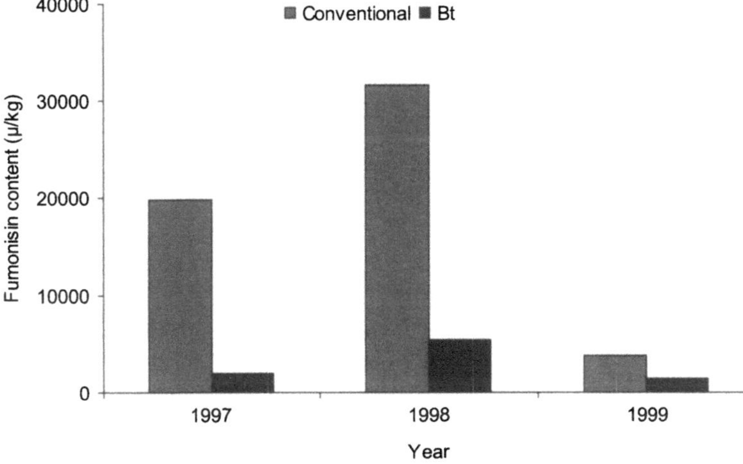

Figure 16.2. Fumonisin (µ/kg) content of maize grain from conventional and insect protected Bt maize grain produced in Italy (Source: Pietri and Piva, 2000).

Table 16.1. Comparison between performance of pigs fed diets containing either conventional or insect protected *Bt* maize on the (Source: Piva *et al.,*.2001a).

Mean values from two locations in Italy	*Maize grain*		*Significance*
	Conventional	*Bt maize*	
Daily gain (g/day)	375	396	P<0.05
Final weight (kg)	22.0	22.6	P<0.05

The Food and Drug Administration in the USA has estimated that the mean economic annual cost of crop losses from mycotoxins including aflatoxins, fumonisins, and deoxynevalenol is US$932 million (http://www.cast-science.org), which highlights the importance of new technologies to control these losses.

The reduction or removal of anti-nutritional factors or toxicants in crops will benefit animals and producers through improved digestibility and nutrient utilization. Biotechnology is the best tool currently available to plant scientists to reduce or eliminate undesirable components of feed ingredients such as erucic acid and glucosinolates in rapeseeds; trypsin inhibitor, lectin, raffinose and stachyose in soybean; tannin in sorghum; gossypol in cotton; and protease inhibitors, phytohemagglutinins, and cyanogens in legumes, in a safer and timely fashion (Hartnell, 2002).

Improved nutritional characteristics of available feed resources

Improving not only the quantity, but also the quality, of available feed resources used in animal production will be essential if the increased demands for livestock products are to be met. Although many different crops and co-products are used in livestock production systems, only a limited number form the main feed resources. On a global basis grass is the most widely used forage resource, but maize silage is also used extensively and is considered one of the best high-energy forage resources available for ruminants. While forages can provide more than 50% of ruminant diets, only limited work has been conducted to date to produce GM forages with enhanced nutritional characteristics. However, it should be noted that conventional breeding techniques have produced brown midrib mutant maize varieties in which structural carbohydrates have been modified to improve digestibility when compared with conventional varieties (Oba and Allen, 2000). The result is that animal performance has been improved. In developing countries crop residues such as maize stover, wheat and rice straw are important forage resources for ruminant livestock. It is recognised that such feeds are of poor quality both in terms of protein and energy content and that this is a serious limitation to enhanced livestock production. In such circumstances there is an urgent need for forages to be modified to contain more digestible fibre components and work to this end is currently in progress at a number of institutions including Purdue University (Vermerris, personal communication). In addition, the development of effective enzymes and rumen bacteria will improve efficiency of fibre digestion and increase animal performance.

Maize grain is often the preferred energy supplement for livestock production systems and soyabean meal is often the preferred protein supplement. Most plant proteins are deficient in at least one essential amino acid, with cereals being generally low in lysine and legumes being low in sulfur containing amino acids such as methionine and cysteine. Modern biotechnology is aiming to improve not only protein content but also the amino acid profile of a wide range of crops which will not only have the potential to enhance animal production but could have a profound effect on human nutrition, where many of these crops are staple diets for the population. The amino acid profile of both maize (Figure 16.3) and soyabean has been modified and the use of such constructs could remove the need for supplementation of specific amino acids, which is often needed to avoid dietary deficiencies (Falco *et al.*, 1995). Thus crops can be enhanced to better match the nutritional needs of ruminant and monogastric livestock. Such developments will reduce the environmental impact of agricultural production systems. Increasing nutrient density of livestock diets by increased non-structural carbohydrates such as starch and oil content (O'Quinn *et al.*, 2000) can also be an important route to improving productivity, efficiency of nutrient utilization and reducing environmental impact.

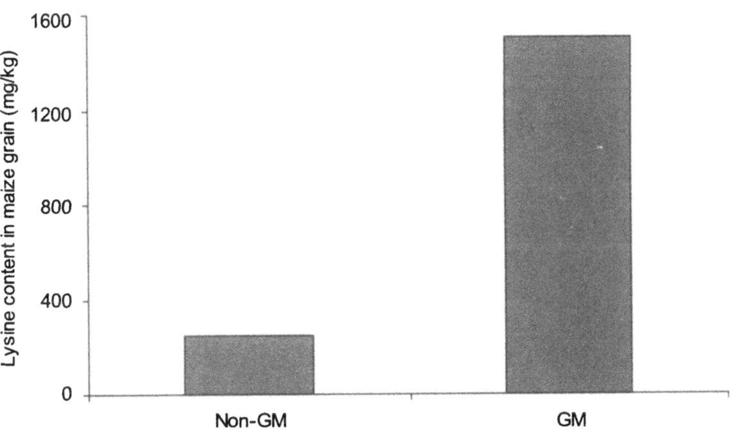

Figure 16.3. Improved protein quality in maize grain (Source: Agricultural Biotechnology in Europe, 2003).

Improved quality of animal products resulting from the use of GM feed ingredients

The consumption of vegetable oils accounts for approximately 20% of total energy intake in industrialized countries. As knowledge increases about their relative health effects, greater emphasis is now being placed on the quality of oils and fats in the human diet (Williams, 2000). Thus the degree of saturation of fats, fatty acid composition, and stereo configuration of individual fatty acids are all targets for genetic modification (Kleter *et al.*, 2002). These modifications are being implemented in a number of crops including, but not limited to, soyabean, maize, canola, cotton, and sunflower. Such modifications will have a marked effect on the fatty acid composition of livestock products (Kalscheur *et al.*, 1997). In addition work is in progress to lessen the danger of outbreaks of food borne diseases during slaughter and processing, as it is now possible through GM technology to make specific antibodies to be produced in crops such as soyabenas that could be fed to livestock prior to slaughter to reduce outbreaks of food-borne diseases.

Improved animal health

In both developed and developing countries, ruminants and monogastrics are susceptible to a wide range of diseases that are responsible for significant reductions in output of milk, meat and eggs. Disease prevention often entails vaccination that can be costly and difficult to achieve effectively and efficiently. Crops are being modified to act as an edible vaccine that can be stored and distributed as seeds, tubers, or fruits. For example workers at the Indian Institute of Science have developed an oral candidate vaccine for rinderpest from genetically modified crops (Khandelwal *et al.*, 2003a, b,). In addition plants have been modified to produce vaccines against a variety of other animal diseases such as rhinovirus, foot and mouth and avian influenza.

Alleviation of abiotic stresses

Many millions of the poorest people in the world live in marginal areas subjected to low and erratic rainfall, extremes of soils conditions, extremes of temperatures and salinity. James (2002) has estimated that 1.5 billion ha of the worlds' arable land suffers from drought. The Nuffield Council for Bioethics (2003) has recently produced a discussion document on the use of genetically modified crops in developing countries including its potential to ameliorate abiotic stress, which can have a marked effect on the supply of both feed for livestock production and also food for human consumption. In it they refer to work at Cornell University, which has successfully produced and tested GM rice that maintained yields when subjected to cold, drought and salinity stresses. Yield increases of 20% are predicted and it is hoped that similar traits can be introduced into other key food and feed crops. Work is also in progress to modify plants to withstand aluminium toxicity, which results from acid soils that are widespread in many developing countries, with Mexico as a prime example (de la Fuente, J.M. *et al.*, 1997). In addition, work at Oklahoma State University has focussed on producing genetically modified wheat with enhanced resistance to water stress and salinity (Abebe *et al.*, 2003). Although these new transgenic crop varieties hold huge potential for large areas of the world where agriculture is practised on marginal land, it is difficult to predict when their use would become commercially viable. The potential significance of GM technology for ameliorating abiotic stress for agricultural in developing countries is shown by the fact that work is being conducted in Argentina, Bangladesh, Bolivia, Brazil, Chile, China, Egypt, India, Indonesia, Morocco, Mexico, Pakistan, Thailand and South Africa on an extensive range of crops (http://wwwfao.org/biotech/inventory).

Conclusions

The paper illustrates how the introduction of first generation GM crops, modified for agronomic input traits such as herbicide tolerance and insect protection, has resulted in significant environmental benefits that will help in development of more sustainable agricultural production systems, through the reduction of inputs such as pesticides and diesel. In addition it has shown how feed resources are being modified for improved nutritional characteristics that will offer significant advantages to livestock producers, not only in terms of increased production and improved efficiency but also in providing food products with improved quality and nutritional characteristics. Additionally, the development of crops able to grow in areas subject to abiotic stresses is anticipated to have a marked effect on the supply of feed for both livestock and human consumption. The potential to provide better animal health through the use of crops genetically modified to act as edible vaccines offers agriculturalists great promise. The importance of improved livestock production should not be underestimated recognising the major trend in developing countries towards an increase in meat and milk consumption with corresponding demands on increased cereal production and less pollution from the inevitable waste products generated. These conflicting objectives are likely to be dealt with most effectively by increasing the efficiency of food and feed production in a sustainable and environmentally friendly way using the opportunities created by GM technology.

References

Abebe, A.C., Guenzi, A.C., Martin, B. and Cushman, J.C. (2003). Tolerance of mannitol-accumulating transgenic wheat to water stress and salinity. *Plant Physiology* 131: 1748-1755.

Agricultural Biotechnology in Europe (2003). *Future developments in biotechnology*. Issue Paper 6. pp. 1-9.

American Soybean Association (2001). Homepage, accessed: 2003, ASA. http://www.asa-europe.org/index.shtml

Barnes, R.L. (2000). Why the American Soybean Association supports transgenic soybean. *Pest Management Science* 56:580-583.

Canola Council of Canada (2001). *Impact of transgenic Canola on growers, industry and environment*. http://www.canola-council.org

Carpenter, J.E., Felsot, A., Goode, T. Hammig, M., Onstad, D., and Sankula, S. (2002). *Comparative environmental impacts of biotechnology-derived and traditional soybean, corn and cotton crops*. Council for Agricultural Science and Technology, Ames Iowa, USA. http://www.cast-science.org

Coyette, B., Tencalla, F., Brants, I., Fichet, Y. and Rouchouze, D. (2002). Effect of introducing glyphosate-tolerant sugar beet on pesticide usage in Europe. *Pesticide Outlook* October, 219: 219-223.

Clark J.H., Ipharraguerre I.R. (2000). Livestock performance: feeding biotech crops. *Journal of Dairy Science* 84 (E. Suppl.):E9-18.

Cunningham, E.P. (2003). Sustainable breeding objectives in developing countries. *Proceedings of the British Society of Animal Science* 2003, pp. 209.

Delgado, C., Rosegrant, M.W. and Meijer, S. (2001). Livestock to 2020. The revolution continues. *Annual Meeting International Trade Consortium*, Auckland, New Zealand, 38 pp.

Falco, S., Guida, T., Locke, M., Mauvais, J., Saunders, C., Ward T., and Weber, P. (1995). Transgenic canola and soybean seeds with increased lysine. *Bio/Technology* 13: 577-582.

Food and Agriculture Organisation of the United Nations (FAO), *FAOSTAT DATABASE 1967-1994*. http://www.faostat.fao.org

Food and Agriculture Organization of the United Nations (FAO), Rome. (1993). Field measurement of soil erosion and runoff. Report by N. W. Hudson, Silsoe Associates Ampthill, Bedford United Kingdom

Fuente, J.M. de la, Ramirez-Rodriguez, V., Cabrera-Ponce, J.L. and Herrera-Estrella, L. (1997). Aluminium tolerance in transgenic plants by alteration of citrate synthesis. *Science* 276 (5318): 1566-1568.

Gardiner, P. and Devendra, C. (ed.) (1995). *Global agenda for livestock research*. Proceedings of a Consultation held at ILRI, Nairobi, Kenya, 18-20 January, 1995. International Livestock Research Institute (ILRI), Nairobi, Kenya.

Giannessi, L.P., Silvers, C.S., Sankula, S., and Carpenter, J.E. (2002). *Plant biotechnology: Current and potential impact for improving pest management in US agriculture. An analysis of 40 case studies*. National Center for Food and Agricultural Policy. Washington DC 20036 USA. http://www.ncfap.org

Hartnell, G.F. (2004). Using biotechnology for the production and enhancement of livestock feed. In: *Dairying: using science to meet consumers' needs*. Edited by Kebreab, E. Mills, J, and Beever, D. British Society of Animal Science Occasional Publication no. 29 (In press).

Ismael, Y., Bennett, R.M., and Morse, S. (2002). Bt cotton, pesticides, labour and health: A case study of smallholder farmers in the Makhatini Flats, Republic of South Africa. In: *6th International conference on agricultural biotechnology: New avenues for the production, consumption and technology transfer*, Ravello, Italy, pp. 6-8.

James, C. (2002). Review: Global status of commercialised transgenic crops: 2002. *ISAAA Briefs* No. 27 International Service for the Acqutision of Agricultural Applicationa (ISAAA), Ithaca NY, USA.

Kalscheur, K.F., Teter, B.B., Piperova, L.S. and Erdman, R.A. (1997). Effect of fat source on duodenal flow of *trans*-C18:1 fatty acids and milk fat production in dairy cows. *Journal of Dairy Science* 80: 2115-2126.

Kent, L. and Moniem, M. (2002). *The potential benefits of genetically engineered crops in Egypt: How much can they reduce pesticide use and increase farmers' income*. Agricultural Policy Reform Program, Reform Design and Implementation Unit, Ministry of Agriculture, Cairo, Egypt, pp. 1-34.

Khandelwal, A., Sita, L.S. and Shaila, M.S. (2003a). Expression of hemagglutinin protein of rinderpest virus in transgenic tobacco and immunogenicityof plant derived protein in a mouse model. *Virology* 308: 207-215.

Khandelwal, A., Sita, L.S. and Shaila, M.S. (2003b). Oral immunization of cattle with hemaggultinin protein of rinderpest virus expressed in transgenic peanut induces specific immune responses. *Vaccine* 3812: 1-8.

Kleter, G.A., Noordam, M.Y., Kok, E.J., and Kuiper, H.A. (2001). *New developments in crop plant biotechnology and their possible implications for food product safety*. Report 2000.004. State Institute for Quality Control of Agricultural Products, Wageningen, The Netherlands. http://www.rikilt.wageningen-ur.nl/news/biotechnology.

May, M. (2003). Economic consequences for UK farmers of growing GM herbicide tolerant sugar beet. *Annals of Applied Biology* 142: 41-48.

Ministry of Agriculture, Fisheries and Food, (1999). *Towards sustainable agriculture*. London MAFF Publications. http://www.maff.gov.uk

Mgheni, M., Mukhebi, A. W., Setshwaelo, L. L. Tsiresy, R. Nyathi P. Osuji P. and Kategile J. A. (1993). Synthesis of constraints to livestock research and development and recommendations. In: *Future of livestock industries in East and Southern Africa*. Edited by.Kategile, J.A. and Mubi, S. Proceedings of a workshop. International Livestock Centre for Africa (ILCA), Addis Ababa, pp. 219-223.

Munkvold, G.P., Hellmich, R.L. (1999). Comparison of fumonisin concentrations in kernels of transgenic Bt maize hybrids and non transgenic hybrids. *Plant Disease* 83: 130-138.

Nuffield Council on Bioethics (2003). *The use of genetically modified crops in developing countries*, pp. 1-164.

O'Quinn, P., Nelssen, J., Goodband, R., Knabe, D., Woodworth, J., Tokach, M., & Lohrmann, T. (2000). Nutritional value of a genetically improved high-lysine, high-oil corn for young pigs. *Journal of Animal Science* 78: 2144-2149.

Oba, M., and Allen, M.S. (2000). Effects of brown midrib 3 mutation in corn silage on productivity of dairy cows fed two concentrations of dietary neutral detergent fiber: 3. Digestibility and microbial efficiency. *Journal of Dairy Science* 83: 1350-1358.

Oerke, E.C. and Dehne, H.W. (1997). Global crop production and efficacy of crop protection: current situation and future trends. *European Journal of Plant Pathology* **103**: 203-215.

Paoletti, M.G., and Pimentel, D.S., (2000). Environmental risks of pesticides versus genetic engineering for agricultural pest control. *Journal Agricultural Environmental Ethics* **12**: 279-303.

Phipps, R.H. and Park, J.R. (2002). Environmental benefits of genetically modified crops: Global and European perspectives on their ability to reduce pesticide use. *Journal of Animal and Feed Sciences* **11**: 1-18.

Pietri, A. and Piva, G. (2000). Occurrence and control of mycotoxins in maize grown in Italy. *Proceedings of the 6th International Feed Production Conference*, Piacenza, 27-28 November 2000, Facolta di Agraria Piacenza, Italy, pp. 226-236.

Piva, G., Morlacchini, M., Pietri, A., Piva, A., and Casadei., G. (2001a). Performance of weaned piglets fed insect-protected (MON 810) or near isogenic corn. *Journal of Animal Science* **79** (Supplement 1):106, Abstract 441.

Piva, G., Morlacchini, M., Pietri, A., Rossi, F., and Prandini, A. (2001b). Growth performance of broilers fed insect-protected (MON 810) or near isogenic control corn. *Poultry Science* **80** (Supplement 1):320, Abstract 1324.

Population Division of the Department of Economic and Social Affairs of the United Nations Secretariat (2003). *World population prospectus: The 2002 revision*. United Nations, New York, USA.

Tillman, D. (1999). Global environmental impacts of agricultural expansion: the need for sustainable and efficient practices. *Proceeding of National Academy of Science USA*. **96**: 5995-6000.

Williams, C.M. (2000). Dietary fatty acids and human health. *Annales de Zootechnique* **49**: 165-180.

17

GM technologies – opportunities and threats of applying GM technology in less developed and developed countries. *'Aladdin's Cave or Pandora's Box?'*

A. Bennett
The Syngenta Foundation for Sustainable Agriculture, WRO 1002.11.66 , CH-4002 Basel, Switzerland.

The context

Globalization describes the many processes by which society is becoming more interconnected and interdependent. It has many dimensions, trade, finance, information, culture, tourism, the spread of pests and diseases, migration, corruption, environmental change and the supply of food. It is here to stay and offers many opportunities for growth and change. The challenge must be, 'can it be made to work for the reduction of poverty, instability and hunger?'

The Food and Agriculture Organisation (FAO) and the Consultative Group on International Agricultural Research (CGIAR) confidently predict that the demand for meat and animal products will increase by 50% between 2000 and 2020 and for eggs by 25% over the same period. The growth in demand will be greatest for pigs and poultry products. There is also a growing demand for fish and other aquatic sources of food.

Increases in demand will be driven by a combination of population increase, urbanization and economic growth. The greatest increases will be in the rapidly growing economies of Asia, particularly in India and China. While local production will meet much of the growing and diversifying needs of consumers it is also inevitable that international trade in animal products will grow. As the wealth of consumers increases their tastes diversify, together with their power to influence markets. As investors or consumers they show a greater interest in sources of food and their wish for information and choice grows. Bovine spongiform encephalitis (BSE) and outbreaks of foot and mouth disease have focused consumer interests on food safety issues. These events have contributed to concerns over the production and use of genetically modified organisms.

It is possible that production subsidies in the rich countries of the world will decline as a result of either a successful outcome to the World Trade Organization (WTO) Doha Round or domestic and economic pressures on public expenditure. These events will have a major impact on global and local markets and patterns of supply and demand in developed and developing countries. However, progress with reforming agricultural policies and subsidies, despite the rhetoric, is painfully slow. In global markets it is still either the efficient or subsidized that succeed.

The increases in demand could open income-generating activities for poor people. There is considerable evidence that the rural poor and landless, especially women, get a higher share of their income from livestock than the better-off rural people. Livestock provide the poor with fertilizer and draught power, an opportunity to exploit common grazing areas, collateral and savings, as well as being an important source of high quality food (Delgado *et al.,* 2003).

The challenge of meeting the growing demand for livestock products will also have to be achieved in the context of increasing demands for a wide range of other economic, social and environmental services derived from our natural resources and the way they are managed. It will have to be done against a background of growing competition for land and water and concerns over food safety, environmental degradation, impact of HIV/AIDs and possible climate change.

The opportunities for increasing the areas of land devoted to livestock production are few and could only be achieved though further erosion of natural habitats, forest, grasslands and swamps, at cost to wildlife and biodiversity.

However it seems inevitable that intensification and increased productivity will be necessary. Intensification in livestock production systems carries with it health and environmental challenges that must be addressed. We cannot ignore the health risks associated with an intensification of livestock production, influenza, zoonoses, parasites and food and water contamination. In addition, in many developing countries productivity is low because of combinations of disease, nutritional and management factors.

Improvements in animal health, welfare, nutrition, management and productivity will be necessary. Advances in genomics and biotechnology could be said to have opened either an *Aladdin's Cave* or *Pandora's Box* of opportunities and tools. Vaccines and diagnostics can play a vital role in improving animal health. Transgenic animals and plants could synthesize an increasing range of proteins and other metabolites with pharmaceutical and industrial value. The GM technologies could increase yields and reduce production risks through imparting drought tolerance, pest and disease resistance and by improving the nutritional value of animal feeds and forage. They could also reduce the presence of pesticide residues in food and water supplies.

However this potential will only be realized if GM technologies are accepted by society as safe and reliable and that they are responsibly used. There are those who have raised the spectres of gene flow, super bugs, allergenicity and long term impacts on health and the environment. These are concerns that cannot be ignored; they must be addressed even if some of them may prove unfounded.

The opportunities - Aladdin's Cave

The supporters of GM technologies argue that they are powerful, precise, innovative, efficient, quicker and likely to produce more profit for producers.

There is good deal of literature on the opportunities for and the potential benefits that biotechnology and genetic modification could bring to animal production and productivity. The enthusiasts argue that GM solutions are powerful, precise, progressive, efficient and quick. However, practical examples are still few. Most have come from investments in the private sector and there are precious few coming from the public sector.

The US Council for Agricultural Science and Technology (CAST, 2003) has recently published an excellent and comprehensive Issues Paper, 'Biotechnology in Animal Agriculture'. Some specific topics (dealing with animal health and the reduction of poverty) are set out in another definitive paper on 'Investing in Animal Health Research to Alleviate Poverty' (Perry *et al.*, 2002).

I do not have time in this paper to go into the opportunities in any detail. They can be grouped as:

- Health
  - Diagnostics to improve and hasten the identification and treatment of pests and diseases
  - Vaccines to provide effective control or possible elimination of diseases. There is also the prospect of introducing vaccines or pharmaceuticals into crops, this is often referred to as *'pharming'*
  - Gene therapy to control inherited predisposition to disease
  - Producing new metabolites and proteins of economic or medicinal value.

  - o Whereas importing countries and retailers, seek to reassure their clientele through more rigorous and expensive regulation and sourcing arrangements
  - o The Biosafety Protocol, which is based on the precautionary principle, requires exporting countries to inform the importing countries of any risks associated with living GMOs and to obtain their advanced informed agreement. The dossiers are developed on local tests and risk assessments. There are as yet few examples of where this gives adequate assurances to importing countries, as was seen recently in the case of food aid in Zambia. The process is very costly and beyond the resources of some organizations and poor countries.

- Environmental impacts
  - o There is concern over gene flow between GMOs, traditional varieties and their wild relative, that this could lead to superbugs, superweeds and contamination of organic production
  - o Many fear that once new genes are released the process is irreversible
  - o The increasing use of high yielding commercial varieties with some form of intellectual property protection varieties could result in the loss of agro-biodiversity
  - o Unforeseen events, 'the surprise factor' 'the BSE effect'.

The use of the precautionary principle lies at the heart of much of the environmental debate and concerns. In applying the precautionary principle attitudes to risk and uncertainty are important. Many people do not accept the argument that GMOs will bring benefits to both producers (through lower costs of production and higher yield), and consumers (though lower prices etc.). Hence they have a very low or zero tolerance to risk. The concept of proportionality in applying the precautionary principle seems to depend very much on the perception of benefits and who are the beneficiaries. There is the added dimension of when using GMOs the risks are not always borne by the beneficiaries, but by others.

The 'unfair' dimension stems from the view that as many of the GM are developed by the private sector and protected by patent, or some other form of intellectual property protection, they are inaccessible to or unaffordable by poor farmers.

The constraints

The debate about the value and use of GMOs is highly polarized.

While it useful to list the opportunities and threats, it is perhaps more useful to review the constraints that currently seem to prevent the resolution of the debate on realization of the opportunities of, or reduction of the threats posed by, GMOs.

The constraints can be grouped as;

- Information or evidence
- Regulation
- Investment
- Delivery or access
- Capacity
- Attitudes and values

Information and evidence

There is a shortage of reliable information and evidence. ICSU (2003) will shortly publish a 'meta-review' of the evidence on the risks and benefits associated with GMOs. It has concluded that there is no evidence of harmful impacts on health. I understand that it will show an uneven distribution of evidence on environmental impacts.

The shortage of reliable and reputable evidence remains the major challenge to moving the debate forward. However there is no shortage of opinion and newsprint.

In the absence of reliable data: who trusts who; what sources do we turn to; who judges; who listens? Who makes a living out of positions, products, production, sales, certification and controversy? Perception is important.

What role should the media play? Is it informing the debate, running campaigns or selling papers and advertising?

Differences within and between countries and communities in the rich world have added confusion to the debate. The differences of view between the US and the EU over the issues of substantial equivalence, choice and labelling are having repercussions elsewhere, especially where the poor get poorer and the hungry starve.

Regulation

Nobody should argue that regulation is not needed; regulation is necessary and must continue to be so! The costs of meeting regulatory requirements, through the development of dossiers of information to be assessed by the regulatory authorities are very high and growing. The expense means that taking innovation to market is beyond all but the biggest or richest countries and companies. This has lead to a consolidation in the industry and a halt or server reduction in the flow of public goods into the market.

At the moment regulations are set at national level. The trend is toward raising the sensitivity of the tests. However there is little progress in getting universal acceptance of national regulatory standards. While countries have the right to choose the appropriate standards of regulation to protect their national populations and environments, international trade would be greatly facilitated by the adoption and international acceptance of national or regional standards.

There remains a question over whether the existing systems of regulation are appropriate or effective, given that there is very little evidence of impact on which to base risk assessments. It might be time to look for alternative systems that would have greater public confidence and which have lower introduction costs.

Capacity

In many developing countries there is insufficient human capital, legislation and physical infrastructure to develop and regulate GM technologies, thereby adding to the growing phenomenon of the 'technology divide'.

The Global Environmental Facility (GEF), the African Agricultural Technology Foundation (AATF), the International Service for the Acquisition of Agricultural Applications (ISAAA), the Consultative Group for International Agricultural Research (CGIAR) and several bilateral and multilateral organisations are supporting capacity building initiatives, but this takes time.

Investment

Who sets the agenda, public good or profit?

In recent years there has been a reduction in public investment in biosciences research and in the fields of genetic modification. This seems to be the result of pressure from those who are concerned about the negative aspects of GM technology, coupled with a perception that these areas of science can be left to the private sector. The private sector on the other hand has focussed its research in those areas where it feels profitable returns might be won and to use intellectual property laws to protect its investments. The costs of regulation and getting products to market before 'the competition', has intensified the focus and pace of work. It has also led inevitably to a greater use of intellectual property protection and the stopping of research that is for public good or too far away from markets that can pay.

I feel that there is a strong case for renewed and increased public investment in these fields.

Attitudes to innovation

While the quality of our lives is heavily based on technological advances, one gets the impression that the standing of science and trust in its motives and standards is at an all-time low ebb. Likewise science in the private sector comes in for even greater mistrust because it is driven by the need to get products to markets quickly and for a financial return on investment.

This requires some serious analysis and discussion.

Where next?

In the richer world we have choice. We have the power through markets, aid and investment to impose our views and values, both locally and internationally. This is one of the features of globalization. I believe that the processes of globalization are here to stay and offer many opportunities to reduce poverty and improve the livelihoods of all. Power should be used responsibly. This power can be used to confuse or bring clarity – but we should also bear in mind that the population of the United Kingdom is only 1 per cent of the world community. We can import most of the goods and service we choose not to produce locally. We can export our dilemmas. These are privileges we should use responsibly.

The growth in demand for livestock products will be in the developing world and theirs will have to be the decision on how they should meet these demands. We can help or hinder.

The predictions are that the growth in demand will be greatest for the products of monogastric animals. This means that there will be a growing market for cereals and grain legumes, and the need to use them more efficiently. We know that there are already technology limitations in these areas. However is the development of new and hopefully better technologies solely the responsibility of scientists? Why have some parts of society turned so strongly against, some of the new technologies and not others?

So do we need scientists, communicators or psychologists?

There is obviously a need for considerably more investment in risk assessment, regulation that is appropriate and the capacity to manage the processes to acceptable standards, but who will make that investment? For the most part it is currently the private sector, but there are many GMs under development in the public sector, such as by the centres of the CGIAR.

It will also require greater international acceptance and trust of regulatory systems and standards. The costs of getting GM technologies into the market place are already higher than more traditional technologies; for many developing countries these costs and capacities needed are beyond the means of their natural institutions. The fact that many of these technologies are emerging from the private sector means they are inspired by market opportunities, but public sector funding has not kept pace and this has created tensions, intellectual property issues and reduction in the availability of GM technologies and risk assessment tools as public goods.

To beat the challenges of rising demand while addressing the problems of poverty, environmental change and instability will require a wide range of skills, technologies and partnerships. Current discord over the use of some technologies and mistrust of the private sector makes constructive discussion difficult. The solution rests in all sides taking risks, build trust and focusing on the real challenge to stability and sustainability, that of reducing poverty.

But above all, there is a need for clearer understanding of, and consensus on, the nature and location of current and future demands and challenges and then some realistic and open analyses of the options and choices for meeting these challenges.

In the end consumers and producers will choose but they must have choice and adequate information on which to base their decisions. That is democracy!

References

Delgado, C.L, Rosengrant, M.W. and Meyer, S. (2003). Livestock to 2020: The next food revolution. *IFPRI 2020 Brief,* International Food Policy Research Institute (IFPRI), Washington DC,USA.

Persley, G. and Lantin M. M. (ed.) (2000). *Agricultural biotechnology and the poor*. Proceedings of an International Conference, Washington, DC – CGIAR.

CAST (2003). *Biotechnology in animal agriculture*. Issues Paper 23. Council for Agricultural Science and Technology, Iowa, USA.

Perry, B.D., Randolph T.F., McDermott, J.J., Sones, K.R. and Thornton, P.K. (2002). *Investing in animal health to alleviate poverty.* International Livestock Research Institute (ILRI), Nairobi, *Kenya*.

ICSU (2003). New genetics, food and agriculture: *Genetic discoveries-societal dilemmas: A meta-review.* Edited by Persley, G. ICSU 2003.

18

Financing the livestock revolution in the developing world: givers and getters

R. T. Wilson
Bartridge House, Umberleigh, Devon EX37 9AS, UK

Introduction

The call for products of animal origin increased dramatically in the Developing World in the last twenty years of the twentieth century. It is expected to double again in the first twenty years of the twenty-first century. The massive demand fuelled a concomitant increase in supply. This combination of events, being increasingly dubbed the "livestock revolution", is certain to have an effect on the global environment. The changes that are occurring can be a force for good: helping to relieve human poverty, hunger and disease and being a driving force in intensification and improved efficiency of livestock production over wide areas. Equally, they can be a force for evil: contributing to environmental degradation, pollution and increased animal malnutrition and disease. The changes cannot, and indeed, should not be stopped but can be harnessed for the first scenario or the one of good. If this is to happen considerable assistance will be required not only in technical but also in financial areas.

Pecuniae obediunt omnia ("all things yield to money" or money makes the world go round). This dixit applies equally to improving livestock production and productivity as to all other aspects of progress. Money is required in large amounts if livestock production and productivity are to be improved and human well-being is to be advanced. There are two sides to the money equation: those who give and those who get. Those who give (or lend) are as diverse a group as those who get (often as credit to be repaid). Sometimes givers are also getters. Not least as givers are livestock owners themselves or their relatives, friends and savings groups. More formally, national governments contribute directly, through production and marketing subsidies, and indirectly, via free animal health services and export credits. Local and international non-government organizations (NGOs) support livestock in various ways. Bilateral agencies in the developed countries also provide dedicated funding for livestock production and health as do some regional development organizations. International and regional development banks and some specialized agencies of the

United Nations system are major sources of funds. Getters, somewhat paradoxically, include several, if not all, of the givers.

Givers

Owners and producers

Livestock are mobile banks for large segments of developing-world populations and are an essentially inflation proof investment. They are also the most important source of protein in the family diet. Most owners of domestic animals are fully aware of these properties and in return are prepared to invest further in their existing assets or make a cash outlay to acquire new ones. One aspect of livestock production that has surprised many development workers, as well as politicians and financiers, is the willingness of owners, and the rapidity with which they have been prepared to take on the responsibility for, in particular, financing, their own animal health interventions. Prior to the financial stringency of the international agencies that began to develop in the 1980s, and the consequent reluctance or inability to finance national Governments in their profligate use of received funds, general veterinary services and vaccinations were usually provided free to livestock owners. Increasingly, under the guise of "sustainable production" owners are encouraged to pay for these activities under "privatization" and "cost recovery" schemes and most do so, after initial reluctance, with little hesitation.

Another major source of investment by owners especially in the arid extensive pastoral areas is in water. As for animal health services, water was often provided by the state in what were perhaps covert attempts to prevent unrest from nomadic pastoralists. It is now much more usual for livestock owners to pay at least the recurrent or operating costs of water although they may receive some subsidy for the initial investment. Even the latter, however, is not difficult to finance internally: in eastern Ethiopia for example a group of 50 pastoral households each with a herd of 50 cattle would require to "pay" only one mature steer per year to provide and maintain a borehole for water for their stock. Purchase of feed and investments in supposedly genetic improvement of native indigenous animals are other forms of investment, as well as the planting of forage crops and increases in labour inputs needed to achieve higher outputs and improved efficiency.

This financial contribution by owners is often ignored or deliberately overlooked in the attempts by financing agencies to achieve media recognition for their own support to livestock development. The fact remains, however, that self-financing by millions of producers injects far more money into livestock production than do the grant and investment projects of what are usually considered the conventional financing sources.

National sources

Animal agriculture accounts for a considerable proportion of the Gross Domestic Product (GDP) of many developing countries. The sector also contributes substantially to foreign exchange earnings through exports of livestock and their products. Livestock convert feed with little or no other value into human food of high and, indeed, essential quality. Domestic animals are critical to survival in drier pastoral areas and are of increasing importance in the intensification process that is a fundamental pillar of the livestock revolution. They support crop agriculture in provision of "intermediate products" including draught power and transport and contribute in some degree to nutrient cycling. In many countries, especially in the Horn of Africa for example, the addition to national GDP from livestock activities is in the range of 15-80%. In some of these economies the value of intermediate products, if fully monetized, would raise the usually calculated value of livestock output by a further 40-50%.

In many cases developing-world Governments have recognized these attributes by offering free animal health services, subsidizing or providing free feeds and supplements and providing border protection against competitive imports. Some if not all of these contributions will be greatly reduced or even descend to zero under donor pressure for "sustainability" and in accord with the precepts of the World Trade Organisation. In summary, Governments provide a considerable amount of investment in the livestock sector. It needs to be noted, however, that budgetary allocations to agriculture in general and livestock production in particular are rarely commensurate with the value added from these activities and even such allocations as are made are often not actually paid out.

Non government organizations (NGOs)

International NGOs have historically drawn their support from individual contributors and philanthropic organizations. Increasingly in the later years of the twentieth century, however, they received funds from Governments who saw the NGO route as a nominal way of reducing their own bureaucratic inputs and devolving implementation to entities supposedly closer to the "grass roots". A phenomenon of the late twentieth century, in part consequent on (often imposed) changes in public/private service balance, right sizing of public sectors and chronic lack of operational funds in public service provision, was the mushroom growth of a multitude of small to tiny local NGOs.

International NGOs usually provide small amounts of finance over rather short terms. Many, however, have grown enormously in size and in social and political influence over time. Some have drifted away from their original constituency. Many are indeed indistinguishable, other than in their constitutions, from large "for profit" entities in their business orientation and the emoluments of their senior staff. Other NGOs remain focussed on their original mandates and not a few of these include a good dose of ethical or religious aspects in the packages they provide to their "beneficiaries". It has become increasingly common for NGOs to act as contracted intermediaries between other types of funding organisms and target populations because of their perceived comparative advantage in working closely with the needy. Thus, for example, they are involved in forming and training groups of producers, or small-scale processors, or act as conduits for management and supply of microcredit. Where NGOs use their own funds for development it is invariably in the form of a grant or gift to the recipients.

Most NGOs are proud of their achievements in development and in many cases have reason to be so. One of their boasts is of their success rate in achieving the immediate objectives of their interventions. This claim is generally true but their interventions are generally on a small scale and involve few farmer beneficiaries. Their long-term impact or sustainability is often questionable once very close supervision is withdrawn.

Bilateral agencies

Bilateral agencies are an important source of financing for developing country livestock improvement. Amongst such agencies are the United Kingdom's Department for International Development (DFID), the German Agency for Technical Cooperation or Deutsche Gesellschaft für Technische Zusammen Arbeit (GTZ), Cooperazione Italiana, the United States Agency for International Development (USAID) and the Swedish International Development Association (SIDA).

Some bilateral agencies provide large amounts of money for development. The main or only source of funds is the public purse that is fuelled by the ordinary taxpayer. National targets (guided by the international community) for the aid bilateral agencies provide are 0.7% of GDP. Very few countries achieve this figure. The constituency of this type of financing is not only target beneficiaries in the developing country (or countries) but also spills over to the concerns of pressure groups in the home country. Bilateral funding is most often in the form of a grant and is touted as being part of the grantors and grantees Official Development Assistance (ODA). The grant is commonly made over a term of 3-5 years, although some countries are seemingly moving to acceptance that livestock development is a long-term activity and adjusting grant periods accordingly. It is not usual, and perhaps more often unusual, for this type of assistance to be entirely "free". It is thus customary with bilateral funding for a substantial proportion of the grant aid to be tied to purchases of equipment from the donor country. There may also be a large component of technical assistance, whereby a national or nationals of the donor country not only act as disciplinary experts but also oversee the financial and administrative aspects of the project.

Multilateral and international agencies

The World Bank, International Fund for Agricultural Development (IFAD) and regional development banks such as the African Development Bank (AfDB), Asian Development Bank (ADB) and Inter-American Development Bank (IDB) are among the major international sources of development funds. The United Nations and its Specialized Agencies are also sources of development funds. Principal among these, or seen as principal, are the Food and Agriculture Organization (FAO) and the International Atomic Energy Agency (IAEA), although these two organizations operate on a small or very small scale in the use of their own funds. The World Food Programme (WFP) also contributes either directly through food-for-work projects or indirectly through allowing Governments to "monetize" food aid. The proceeds of this monetization are then supposed to be used for development purposes including development of livestock although it goes almost without saying that they are not always applied to this end. One former large scale donor, the United Nations Development Programme (UNDP), has moved away from direct support for "disciplinary" projects such as livestock, crops or irrigation to channel its resources to "crosscutting issues" such as gender, democratic governance, crisis prevention and recovery, conflict resolution and the environment. These last types of interventions may be only peripheral and seem indeed to be coming increasingly marginal to livestock. Almost all of the principal organs of international finance operate in different ways in the name of development. Over the years, however, some commonalities have developed.

Funding by these international agencies is of various types and comes in various guises. The agencies themselves are usually "replenished" by their members who are national states or other multilateral organizations. Funding is mainly as a loan to a developing-world Government. The project or programme for which a loan is provided is known as an "investment" and not a "development" project. In theory terminal activities must show good financial returns to producers and long-term economic gains to the recipient country, with a presumption that these gains will repay the loan. Loans are of various

Table 18.2. Percentages of project costs by "expenditure accounts".

Main financing agency	*Country*	*Project type*	*Percentage of total expenditure*							
			Civil works	*Vehicles*	*Equip-ment*	*Train-ing*	*Technical assistance*	*Other*	*Credit*	*Recurrent*
International Bank 1	Mozambique	Smallholder Irrigation	33.0	3.0	19.0	11.0	4.0	8.0	9.0	14.0
International Bank 1	Mozambique	Family Livestock	0.2	17.3	23.3	16.4	17.8	4.2	0.0	11.0
International Bank 1	Sierra Leone	Livestock Development	12.0	10.0	5.0	10.0	20.0	0.0	19.0	23.0
International Bank 1	Nepal	Uplands Development	51.0	1.0	4.0	8.0	13.0	0.0	7.0	16.0
International bank 1	Egypt	Delta Rural Development	18.0	1.0	3.0	7.0	4.0	22.0	27.0	16.0
International Bank 1	Turkey	Eastern Anatolia Development	30.0	0.4	2.1	5.3	4.4	0.0	42.7	5.4
International Bank 1	Turkey	Central Anatolia Development	24.0	1.0	0.0	8.0	4.0	45.0	0.0	12.0
International Bank 1	Armenia	Agricultural Services	39.0	4.0	5.0	2.0	4.0	2.0	26.0	19.0
International Bank 1	Bosnia	Livestock/Rural Finance	0.0	1.0	7.0	3.0	9.0	0.0	67.0	13.0
International Bank 1	Vietnam	Agricultural Resources Conservation and Development	40.0	5.0	3.0	5.0	7.0	2.0	29.0	9.0
Regional Organism	Horn of Africa	Regional Livestock	6.6	6.0						
Regional Bank	Ethiopia	National Livestock	13.7	13.1	50.5	6.3	1.6	2.0	0.0	18.0
Regional Bank	Ghana	Livestock Development	3.0	13.0	19.0	29.0	6.0	0.0	7.0	23.0
Regional Bank	Kenya	National Livestock	11.0	3.0	15.0	21.0	16.0	21.0	0.0	14.0
Regional Bank	Zimbabwe	Small Animal Production	0.0	22.0	4.0	30.0	1.0	0.0	17.0	27.0
Regional Bank	Zimbabwe	Smallholder Dairy	0.0	6.0	0.0	22.0	1.0	0.0	62.0	9.0
Regional Bank	Zimbabwe	Beef and Draught Power	0.0	4.0	1.0	68.0	9.0	5.0	5.0	8.0

of any funding (Table 18.2), are subject to international competitive bidding (ICB) with supply contracts awarded to the lowest bidder. Control of the project by the loan agency, once more in theory, is through regular supervision missions, mid-term reviews and project completion reports, but here again recipient countries are reluctant for these activities to be financed from the loan itself.

The international financing agencies and regional banks also provide grant funding additional to that for direct technical assistance. This is usually, however, a small part of the total portfolio. The World Bank, for example, provides grants for agricultural reform and especially institutional, policy and land tenure reform under its Structural Adjustment Programmes and Agricultural Sector Support Programmes. The IFAD has a Technical Assistance Grant activity (which is less than 10% of its total portfolio) to support research which is then mainly carried out by international research centres (such as the International Livestock Research Institute based in Nairobi in connection with livestock). A condition of such a grant is usually that research is carried out in collaboration with national organizations (National Agricultural Research Systems, NARS). The research for which funding is provided is mainly intended to be transferable directly to IFAD's loan projects but it is also not unusual for these projects themselves to have a small grant element for research that they can contract out to a suitable organization.

Getters

Owners and producers

It might be expected that livestock owners would be the principal beneficiaries of the funding that is available for development. Regrettably this is not the actual case. Rather they invest much of their own current cash and savings in their activities. In return they take most of the risks involved in livestock production and, as in the developed world, very often receive little recompense for their investment and industry. Owners do benefit from financial support directly and indirectly but there are often considerable costs associated with this. As can be seen from Table 18.2 they receive a variety of benefits but many, such as training, can be rather nebulous. In other cases the cost of the assistance they receive, compared to the costs of those higher up the development chain, is high. A case in point is that of credit in internationally financed projects. Whereas the "Executing Agency" (the borrower of the loan and almost invariably the national Government) receives credit at very favourable rates there is an almost exponential increase in the interest rate until it reaches the farmer stratum. Such usurious levels often result in high rates of default in repayments of both the capital of the loan and its accrued interest.

National Governments

National Governments are probably the recipients of the greatest benefits of attempts to improve livestock production and productivity and of actual improvements. Multiple sources provide these benefits but the two major ones are increases in revenue from greater livestock output and incomings from grants and loans from financing agencies. Revenue from higher livestock output is mainly from higher taxes at first point of sale (very few Governments now apply the head taxes that were always contentious and a source of constant friction between administrations and stock keepers) supplemented by more earnings from slaughter charges. Other additional income arises from further taxes such as the Value Added Tax (VAT) and Sales Taxes imposed at various rungs of the supply chain. A major source of the much sought-after income that is foreign exchange from export earnings which is itself supplemented by port loading and handling charges derives from the international trade in livestock.

As can be seen from Table 18.2, major receipts for national Governments from bilateral and multilateral assistance are the provision of extensive new infrastructure and mountains of equipment. Among the former are staff houses, laboratories, livestock handling and market facilities, inland and at export ports, and access roads. Equipment includes (usually expensive and seldom necessary 4-wheel drive) vehicles, laboratory instruments and computers and peripherals. In at least one (notoriously failed) project in Sudan huge amounts of money were spent on railway locomotives and rolling stock. Traders preferred not to use these because of the cost and because an important factor in the market chain was the 3-month rainy season trek from producing area to terminal market, which allowed cattle to fatten over the vast distance covered and not destroy resources in a single location or require expensive and scarce feed during a fattening period in lots or holding grounds.

Individuals in developing countries

People in the public and, increasingly, in the private sector who are not producers or owners of livestock are amongst the most prominent direct beneficiaries of livestock development. Human capacity building is the name for this process. Extension agents are provided with training in the activities that a project is to promote. These people are then expected to pass on their newly acquired skills to the "direct beneficiaries" after this "training of trainers". Study tours are another way that trainers (rarely producers) are provided with the skills they are supposed to have to assist livestock owners to become more efficient. It is not rare for senior civil servants and even ministers to be provided with such opportunities to enhance their capacity.

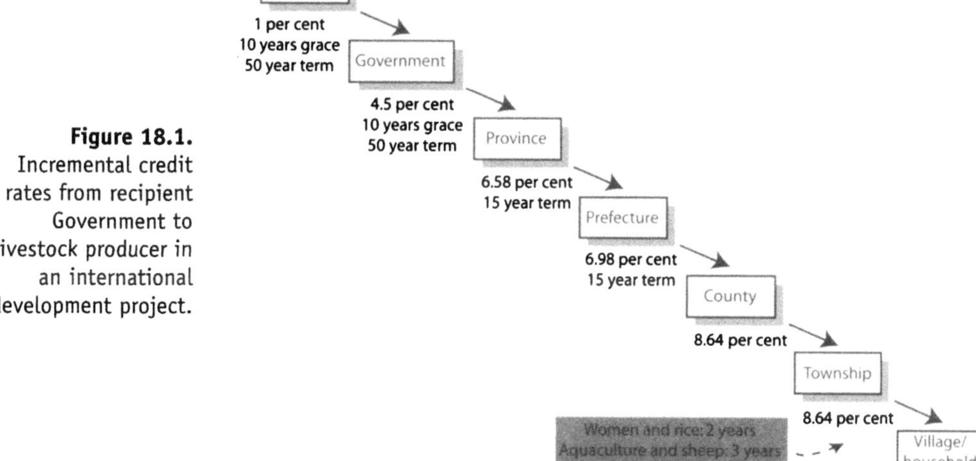

Figure 18.1. Incremental credit rates from recipient Government to livestock producer in an international development project.

Recipients in bilateral agency countries

As already indicated, but increasingly denied, much bilateral aid is "tied" in some way. This may include a long study period of up to three years by a large team of individual experts contracted to a consulting firm incorporated in the donor country (although in the European Union nationals and commercial firms based in any of the countries of the Union are supposedly considered equally in the award of contracts). Subsequent implementation of the project (if one does materialize from the study) has the same characteristic structure. Additional amounts of that which is supposed to be official development assistance are channelled to home country manufacturers.

In many respects some international or multilaterally financed projects mirror those that are funded from bilateral sources. The main difference here is that "procurement", including that of international technical assistance personnel as well as of equipment, should be done through ICB procedures.

International agencies

Some perceived givers are in effect net getters of aid contributions. One international organization whose mandate includes "the alleviation of poverty and hunger by promoting agricultural development, the pursuit of food security by providing all people at all times with access to the food they need for an active and healthy life, and the continuance of direct development assistance" is a case in point. In a project designed in 2003 it proposed a certification system for exports of livestock and products of animal origin and was seeking US$ 207 000 in funding from a multilateral donor for its implementation. Of this amount 56% was for personnel (43% for international "experts", 9% for local experts and 4% for administrative assistance), 19% for travel of these personnel, another 19% for general operating costs, 5% for direct operating costs and 4% for board meetings. Less than 10% was to go to training field workers (the only direct local beneficiaries) in meat hygiene and in order for the project to go ahead the local private sector (that is, traders and exporters) was required to put up the money to train another 10 meat inspectors.

International money is channelled back into the international organizations in other ways. One route is through the system of "supervision" whereby one entity is appointed as the financial and technical overseer of a project financed by another. It might seem more appropriate in the climate of privatization that is ardently promoted by the international donor community to cede this activity to a private sector firm. This option might also lead to greater "transparency" in the success or otherwise of project implementation and help to overcome the opaqueness of many bland supervision and project completion reports. Transparency has been further fogged since the late 1990s by allowing one UN agency to act as an agent in the ICB process through which system even more money can be diverted back into the financing system from the poorer relations in the development chain.

Discussion

Producers provide most of their own financing for livestock development although this is usually accorded a low profile in comparison to that provided by national Governments and the bilateral and multilateral sources. Similarly whereas the main "target group" is usually touted as the livestock producer the physically quantifiable benefits received by that class of beneficiary tend to be small. Even the indirect benefits of improved livestock performance can often be only tenuously linked to external support in improved capacity of the public services to provide support, as for example in delivery of animal health services or production extension advice.

Whether provided as grant or loan, funding by external agencies (with the possible exception of NGOs) is seldom seen as solely altruistic nor yet to increase livestock production efficiency *per se*. It is used by donor or lessor to further the ends, or "mandate" and "mission", of the agency itself. Alleviation of poverty and food security are highest among these ideals (one international bank's mission is to "enable the poor to overcome their poverty"). Policy reform is another favourite intervention. For many years "gender" (usually meaning women in development jargon) has been a major issue. Projects that favour women, either by directing financial and physical benefits to them or that reduce their workload, are thus usually regarded with particular benevolence by the funding community. Sustainability, the environment and natural resources have become important words in the development lexicon. In not a few cases sustainability would be better expressed as continuity, which more closely reflects the hope that a project will continue in its activities after the end of the 'x-year' project funding period. Even continuity, however, has proved an elusive goal. A guiding principle in project design is to keep recurrent costs to a minimum in order to reduce the burden of continuity after project completion, with a corollary that recurrent or operating costs even during the project life are to be paid by the recipient. Funding agencies frequently ignore this last stricture as many projects would not operate if the "recurrent costs = local costs" principle was adhered to strictly. The consequences are to be seen everywhere in dilapidated buildings, degraded roads and the dozens of vehicle carcasses sporting project logos that adorn the compounds of Ministries of Agriculture and Departments of Veterinary Services. Similarly many, indeed almost all, projects designed or implemented since the early 1990s have a component described as natural resources conservation or management. These components may be little more than those which earlier, in the context of livestock production, might have been included under the rubric of pasture improvement or forage production but these latter terms at the start of the twenty-first century are not seen to have the same high profile.

A further prominent agenda item is institutional strengthening which in its most common manifestation comprises the supply of vehicles and equipment to NARS and other Government or parastatal bodies. Institutions can, it seems, also be strengthened by collaboration with international research centres or advanced centres of excellence in the developed world. Directing funds at these second kinds of organizations does of course bleed them away from entities closer to the problems. For the international agencies all of these activities combined are intended to create an "enabling environment" that will in some remote way improve livestock production.

Enhancement of human resource capacity in many cases cannot be distinguished from training. It may be intricately woven into the fabric of institutional strengthening as a further and, indeed, real necessity in most projects. Many scientists and administrators in developing or recipient countries view capacity enhancement as a chance to travel abroad in pursuit of a postgraduate degree (sometimes in a not entirely relevant discipline and from which they may fail to return to their original post). Funding agencies now regard this route to development with less favour than hitherto. At least one regional bank, and other agencies are moving in the same direction, will not provide loan or grant funding to this kind of personalized capacity building. More common forms of building human capital that have developed since the mid 1990s are relevant short training courses in the activities of the project or in-service training for extension workers. In investment projects training of farmers in aspects of production that are being promoted, such as cross-breeding of dairy cows, small-scale intensive poultry keeping, use of modern hives for honey production or post production value-added processing may be prerequisites to obtaining project sponsored credit with which to set up the modern or improved enterprise.

Empowerment is integral to human capacity but is often treated as a separate component. The poorest of the poor, women and marginalized groups such as landless livestock owners or ethnic minorities are targets of empowerment. The process, in which local and international NGOs are usually the main intermediaries because of their believed skills and experience, often includes setting up producer and marketing groups and associations, training of such groups in simple management and accounting, providing basic or functional literacy skills, assisting in the formulation of microproject documents, facilitating access to credit and creating awareness of fundamental human rights so that poorer people are better fitted to do battle with local, provincial and national authorities. In its 2001 Rural Poverty Report, one of the international lending agencies expands on its simple mission statement and goes on to say: "Poverty reduction is not something that Governments, development institutions or NGOs can do for the poor. They can forge relationships and help promote the conditions in which the poor can use their own talents and skills to work their own way out of poverty."

Most of the activities that are described in the preceding paragraph and that are fostered by the funding agencies are generally accepted without much comment by recipient countries (except perhaps sustainability, environment and, especially, empowerment by some recidivist administrations) but other propositions are less likely to be. Grant funding can be used in particular by donors to try and force unpalatable activities on reluctant or recalcitrant countries. One example is that of the research institutions of the former Soviet Union. Whereas almost all land has been privatized and allotted to small or even tiny holdings the behemoth centres that survive still target their research at large state farms (that failed even before the fall of Socialism), although what is required is adaptive research that will benefit the new class of smallholders. Privatization and full cost recovery are other unpalatable items for some governments and even for some practitioners: animal health services are the subject of both these aspects of "development". Where they have not been rejected outright implementation has almost always been slow and partial which most probably illustrates the very different mind sets of those who have money and need to share it and those who do not have it but are in want of it.

Conclusions

Finance for livestock operations is necessary to improved efficiency. Livestock owners themselves are the main source of investment. National Governments add to (and subtract from) livestock development in various ways. Several external funding agencies operating in various ways are involved in livestock development. A large part of the finance other than that deriving from owners, by whomsoever and in whatever way it is provided, is not devoted to directly productive activities. External funding, or the promise of it, is used to promote what the financing agencies see as necessary to make the world a better place to live in and to improve human well-being. The new paradigms include institutional and policy contexts, partnerships, local "ownership" and more effective monitoring and evaluation. In these goals the givers are not always at one with the getters but *Pecunia non olet* ("money doesn't smell" or don't look a gift horse in the mouth) and getters often spend the money they receive (or don't spend it) on activities other than those agreed. The consequences can always be discussed at a later date. These differences in perceptions may account, in no small part, for the failure of many development projects to achieve their goals and for the slow pace of development in general.

Acknowledgements

The information provided in this paper derives from the author's experience as a practising rancher, development worker, research scientist and consultant to bilateral and international organizations. Much of the original data comes under the heading of "confidential" and it is not, therefore, possible to quote the sources directly. I am, nonetheless, grateful for the opportunities that have been given me to participate in the attempts to improve livestock production in the developing world and, perhaps more personally, to be a net "getter" in the process.

19

Knowledge management and dissemination for livestock production: global opportunities and local constraints

C.J. Garforth
School of Agriculture, Policy and Development, The University of Reading, PO Box 237, Reading, RG6 6AR, UK

Coping with global knowledge expansion

Science produces new knowledge relevant to agricultural production at a breathtaking pace. Despite concerns over the limited resources committed to publicly funded research, the number of articles in scientific journals and of papers and posters presented at scientific conferences continues to increase year on year. Increasing amounts of knowledge are also being generated in the commercial sector, some protected by commercial considerations but much of it contributing to the general stock of publicly available knowledge. Yet individuals have a limited capacity to attend to and take in new information. We all face the frustration of not having enough time to read the relevant materials that pass across our desks, let alone enough time to search the literature to update our knowledge. Scientists have developed mechanisms for handling the sheer quantity of communication of new knowledge among themselves and for dealing with questions of quality control and assurance. These mechanisms include specialisation within narrower disciplines, accepted protocols for experimental design and statistical analysis, peer review of outputs, publication of scientific journals, networks and conferences.

Governments have also created mechanisms for ensuring that policies and policy making are informed by new knowledge generated by science. These include the appointment of scientific advisers, the establishment of science units within government departments and the commissioning of research to provide knowledge that will inform policy decisions. At an international level, FAO among others is investing in the creation and maintenance of searchable on-line databases as a means for policy makers and advisers across the world to identify relevant knowledge and information, taking advantage of new technologies and communications infrastructure to allow users to access relevant information as and when they need it.

We therefore have procedures in place to facilitate the sharing of knowledge among scientists and with policy makers. We all have concerns over the effectiveness and efficiency of these and we should certainly not be complacent that there is no room for improvement. There remain more serious concerns, however, over the effectiveness with which new knowledge – and indeed "old" knowledge – is being shared with livestock producers. There is a widespread perception that knowledge generated by publicly funded research is not being taken up by farmers. In the UK, the Department for Environment, Food and Rural Affairs (DEFRA) has recently sponsored a set of Knowledge Transfer Initiatives, is developing a Knowledge Transfer Strategy and is funding research into knowledge transfer processes (Garforth *et al.*, 2003). On the international front, the UK Department for International Development (DFID) is pressing managers of its various research programmes to demonstrate that their research outputs are being taken up by farmers and "target institutions" and are having a positive impact on the livelihoods of farmers and others in developing countries – particularly those who come within the scope of the International Development Target of substantially reducing the numbers living in poverty by 2015. International institutes are increasing the resources they devote to technology transfer and dissemination. These concerns and initiatives raise considerable challenges, both conceptual and methodological.

Knowledge transfer or communication?

There is a considerable literature on the relationship between information and knowledge. Some argue that they are the same thing. Others argue that knowledge remains personal to the knower, while information can be exchanged between people: the individual's knowledge is affected by the information to which he or she is exposed (e.g. Röling, 1988). This paper takes a more pragmatic position: the term "knowledge" can be applied to facts and relationships that are widely accepted or can be demonstrated by accepted protocols, as well as to the personal knowledge that is unique to the individual. We communicate that new knowledge, however, by providing information both about its substance and about how it was generated.

The transfer of knowledge to those who make production decisions – farmers, members of farming households, farm managers, farm workers – is often seen as the translation or conversion of knowledge derived from science into "messages" which can be applied on farm. For some knowledge, this model seems to fit: for example, encouraging dairy cattle keepers to follow basic hygiene practices when milking is a logical outcome from research which has generated knowledge about processes of contamination of milk and transmission of infection between cows. It is a message which is relevant to all dairy producers, at all scales of operation and in all circumstances – although how it can best be applied will vary considerably. More generally, however, there are five difficulties with this simple model. First, farm situations differ, in both internal and external characteristics. At the most basic level, animals are individuals with their particular genetic potential, behavioural characteristics and propensity to disease. Farm businesses differ in their mix of enterprises, distance from markets, institutional affiliations, labour and capital availability, investment plans and management priorities. Many decisions therefore require the input of farm-specific data before "science" can deliver an appropriate suggestion. Hence the widespread development of decision support tools, especially in the area of animal nutrition. These differences also raise the question of what balance we should be trying to achieve between "transferring" knowledge of basic principles so that the farmer can decide how to take these into account in the specific circumstances of his or her enterprise, and giving recommendations or advice on specific practices.

Second, "knowledge transfer" is a contested construct. At the level of the farming industry, it seems to make sense: knowledge can be said to have been transferred when a new idea, product or practice, developed and tested through R&D, becomes widely used by farmers. But at the level of the individual who must decide whether or not to use the new idea, product or practice, the concept is more problematic (Garforth and Usher, 1997). Knowledge transfer implies a knowledge-deficit model of communication: I tell you something that you did not know before, which you can therefore simply slot in to fill a gap in your existing knowledge. Communication, however, does not work like this: it is always an interaction between two sets of knowledge. We can pass on information, but the effect of this on what the other person "knows" is unpredictable and may be very different from what we thought we were telling them. New information does not simply fill a gap in what the person already knows, or replace an existing item of knowledge: it interacts with what is already known – perhaps reinforcing what the person already knows (through either accepting or rejecting a new idea), or modifying it. The interpretation of new information is affected by the person's existing knowledge and concepts. Even our simple message on hygiene around milking may be more problematic than at first sight if farmers interpret "hygiene" or "clean" differently from scientists. Knowledge is not just a collection of objective facts that we carry around with us. It includes subjective assessments and beliefs. In the context of a new technology, product or practice, an individual's knowledge includes what he or she believes about its efficacy, relevance, and superiority to alternatives. Information about it will be evaluated against this existing knowledge.

Third, in all but a limited set of circumstances, effective communication is always interactive, more like a conversation than listening to a lecture. Exposure to new information may raise questions in the minds of those who hear it, requiring clarification or further information. In scientific circles, how often have we heard (and voiced) criticisms that a conference programme did not allow sufficient time for questions and discussion of the papers?

The fourth difficulty is a more practical one, relating to quality control and assurance. Science, as indicated earlier, has developed robust procedures for assuring quality in the generation of knowledge. The farmer is in a much worse position: how does he or she assess the validity and reliability of the information they are presented with? This is a recognised problem in relation to physical inputs such as fertiliser and pesticides where adulteration and misrepresentation are not uncommon in many countries (Williamson, 2003). In the absence of recognised quality assurance schemes for information, deciding whether to believe what one hears or to rely on it in making production and management decisions is influenced by the credibility assigned to the perceived source or means of communication.

The fifth point derives from a fundamental difference between science and farming – the degree of fragmentation and specialisation of knowledge. Science has evolved by becoming increasingly specialised. New disciplines are created and existing disciplines are divided as the search for new knowledge pursues its course. New, more specialised journals emerge to communicate the findings of these new disciplines within their own ranks. Farming, too, has become more specialised in many places, but at the very least farmers are managing complete animals and production cycles. In developing countries, livestock enterprises are often integrally linked with crops, as in the hills of Nepal where crop and livestock production depend on one another and are both dependent on access to forest resources. It is the farmer, in the end, who has to integrate new knowledge into production systems on the farm. This includes making judgements on the trade-offs between alternatives and assessing the financial implications of any change in management. This raises questions about where the responsibility lies for mediating between discipline-specific scientific output and integrated decision making on farm.

Some of these difficulties are highlighted in current research on the factors which constrain the uptake of new ideas to improve heat detection in dairy herds in England and on animal health knowledge among landless dairy cattle keepers in India. Scientists know that fertility rates among UK dairy cattle are declining and that this is partly due to some heats going undetected. There are technologies available, based on scientific research, which have been shown to improve heat detection but have not been widely adopted by farmers

(Cooke and Park, 2002). Focus group discussions and a survey of 500 dairy farmers in south-west England, however, suggest that the majority of farmers do not think they have a problem with heat detection, judge the new technologies not to be cost-effective, and feel that their own stockmanship, experience and record keeping, combined with advice from the vet during routine visits, are the most reliable aids to heat detection (Rehman et al., 2003). In a baseline survey for the India study, 890 landless owners of dairy cattle were interviewed to determine their current knowledge about symptoms, causes and treatment of ten common diseases and conditions of dairy cows. While the survey did identify things which respondents did not know – an important first step in developing information and training materials – it also showed elements of their current knowledge which scientists would generally regard as incorrect but with which any new information will inevitably interact (Ramkumar, 2003). Both UK and Indian studies have so far confirmed the high credibility associated with veterinary surgeons.

A further complication of the simplistic model is that new knowledge does not just come from science. Farmers are also finding out new things for themselves as they modify their management and practices in response to changing circumstances. While some of this may have little relevance beyond their own farm, much of it is shared with other farmers locally. Some of it may throw up hypotheses and questions which can only be addressed through formal scientific research. The relationship between science and practice is therefore not linear: it is dynamic and iterative. In terms of knowledge management, one challenge is to ensure that knowledge generated at the farm level which is potentially relevant more widely can be identified and made available; and that any potentially harmful consequences of the application of such knowledge are identified and addressed.

Opportunities for knowledge transfer

We start from two basic realities. First, farmers access information from many different sources. Information about new knowledge and its application can therefore be placed with those sources and channels which are most likely to be accessed by the set(s) of farmers for whom it is relevant. Second, different kinds of knowledge are best learned in different ways. For example, knowledge of how and how frequently to apply a particular acaricide can be learned from reading a label on a bottle (or, for a farmer who cannot read, having it read to them). Learning what non-chemical methods of tick control are effective locally, on the other hand, may only come from being part of a group of livestock owners who agree to carry out experiments in a setting such as a farmer field school (FFS) (Minjauw *et al.*, 2003) or a Monitor Farm (Riddell, 2001). There is now a lot of knowledge around on how (and how not) to communicate different kinds of information and new knowledge effectively.

For livestock health and production, the veterinary surgeon is an obvious route by which knowledge reaches producers and a source which is highly credible among producers because of their actual and perceived expertise. But they are also very busy people. In India, where government vets have a role in providing advice on animal husbandry, very little time is spent specifically on this aspect of their work (Ramkumar, pers. comm.). Most visits to or by the vet will lead to some exchange of information which will have the effect of reinforcing producers' knowledge or exposing them to something they did not know before. Outside the context of these clinical consultations, however, vets can spare little time for more extensive knowledge transfer activities.

Other inter-personal channels are other farmers and service providers. In the context of small scale dairy production in peri-urban India, these latter include professional milkers, local milk vendors and retail outlets which sell concentrates (Ramkumar, 2003). Training for these people, and provision of print material (such as leaflets and posters) for them to hand on to producers, can enhance the flow of information about new knowledge within the farming community.

Beyond inter-personal channels, there is a wide array of media: some of these are useful for providing information in a linear fashion, while others allow various degrees of interaction between knowledge source and user. Print is widely used, from labels on product packages, through leaflets and posters, to booklets and manuals. A lot has been learned about the effectiveness of different formats in different contexts, such as the "Wambui finds out" series of booklets based around characters in a rural household in Kenya which were developed and tested in the context of a project funded by DFID's Livestock Production Programme (e.g. Bain and Tanner, 1999). The commercial sector has researched the communication of product safety information through symbols. As with all media, the main requirement is to design print materials with the audience's context and perceptions in mind – and preferably with their active participation – and to pre-test them before final production and distribution. The grey literature is full of accounts of print material that were interpreted in entirely different ways to that intended by the designer – including the poster in which farmers saw arrows connecting the stages of the lifecycle of the tapeworm as snakes, footpaths or road signs.

Radio has been used for getting information to farmers for over fifty years. Formats vary from talks and interviews, to short spot announcements, magazine programmes and long running dramas.

Radio is still the medium with greatest penetration in rural areas worldwide, though it is beginning to be supplanted by television in some places: a recent survey in peri-urban areas of Pondicherry showed that over 75% of landless livestock producers have a television in their home compared to less than 40% with a radio (Ramkumar, 2003). However television, even more than radio, is a medium of entertainment, and with multiple channels for the audience to choose from it is difficult to ensure exposure of a particular audience to information and educational content related to livestock management. The spread of local FM radio, on the other hand, offers new opportunities for relaying locally relevant information and for engaging the audience in dialogue through phone-in broadcasts and field recordings.

Computer-based information and communication technologies (ICTs) are seen by many as a giant leap forward in people's ability to access the information they need, particularly where Internet access is available. Searchable databases with user-friendly interfaces (touch screens, pictures and audio in addition to text) are a convenient way for users to access knowledge on animal health. An information kiosk being piloted by the Rajiv Gandhi College of Veterinary and Animal Sciences (RAGACOVAS) in Pondicherry, for example, contains information on symptoms, causes, treatment and prevention of major diseases of dairy cattle – the contents of which were based on a participatory appraisal of health knowledge and concerns of landless cattle owners in the area (Ramkumar, 2003). Its touch screen (no mouse or key board), friendly graphics, animations, audio delivery of information and choice of language (Tamil and English) are designed to minimise the difficulty livestock keepers will face in using it. Use and feedback will be monitored, with a view to improving it and installing further units in service centres. There are concerns, however, about the wider utility of such technology based on experience relating to maintenance, reliability of power supplies, updating of information and the cost of both hardware and software. So far, computer-plus-Internet facilities at community information centres have not spread beyond specially funded pilot projects – and many do not survive beyond their pilot phase. For Internet applications, the lack of locally relevant information in local languages is a major drawback, which is compounded by the sheer volume of material, lack of any protocols for quality assurance of Internet content, and the time and skills needed for an efficient search.

Challenges in an era of globalisation

Access: the digital divide and information poverty

Despite continuing rapid developments in communications technology and the global spread of its infrastructure, access to sources and channels of information remains very uneven. Even in high income countries, less than a quarter of the population are Internet users while in South Asia there are less than 120 radios per 1000 population (Table 19.1). In the UK, a recent report by IBM's Local Futures Group (Oakley and Campbell, 2002) warns of a "new underclass of people in rural and remote areas who are being excluded from ... online public services by lack of access to technology". With the commercial sector driving expansion, it is inevitable that those who are already well connected to sources of information will be the first to benefit from new developments, thus increasing the digital divide. This is particularly problematic at a time when producers are being encouraged to compete in global markets, where rapid access to market information (prices, consumer requirements, regulations) is needed. For small-scale producers, it is difficult to compete with the superior access to information and knowledge of larger scale operations. Small-scale vegetable producers in Kenya have already seen a large part of their share in the export market lost to much larger enterprises.

Table 19.1. The digital divide between nations: basic indicators (number per 1,000 people).

Category of country	*Radio sets*	*Televisions*	*PCs*	*Internet users*
Low income	157	85	4	1.9
Middle income	360	279	27	16.8
High income	1289	693	346	212.5
Within the low and middle income category:				
Latin America and the Caribbean	419	272	38	18.8
South Asia	113	71	3	2.2
Sub-Saharan Africa	201	43	8	3.6

Source: World Bank Development Indicators 2001 http://www.worldbank.org/data/databytopic/databytopic.html

Unequal access to information occurs at all levels – within households, within a village, between villages in the same area, between urban and rural areas, as well as between nations. An information mapping study in Eritrea, for example, showed distinct differences between men and women in their access to various sources of information; it also showed that two villages just 10 kilometres apart had vastly different access to external sources of information (Garforth, 2001), with one mainly reliant on a limited set of formal government channels while the other had access to a multiplicity of non-government channels through market links and local organisations.

Improved access to the means of obtaining information for poorer producers requires public investment coupled with regulatory frameworks which create parameters within which commercial providers of communication infrastructure operate. Making effective use of that access may require some form of mediation – for example, a local radio station providing an Internet search facility, or a community information centre which has voluntary or salaried staff to help users find information that they need. The model being developed by communities in twelve Information Villages in Pondicherry with the support of the M S Swaminathan Research Foundation (MSSRF, 2003) is an example of such mediation, enabling people in the villages to access dynamic information (such as daily prices in different markets, sea conditions) and more static information (including technical advice on many aspects of agriculture and rural livelihoods) very quickly from credible sources. This reduces the cost of their information seeking, redresses the balance of power between them and various actors in the social, economic and political system (traders, moneylenders) and improves the quality of their decision making (Rajamohan, 2003).

Quality assurance

Information is often defined as "that which reduces uncertainty". This definition only holds if the person exposed to the information is confident of its veracity and relevance. Faced with conflicting pieces of information, or information which conflicts with existing knowledge, the degree of uncertainty the individual experiences will increase rather than decrease. Where the source is perceived to have a vested interest in their information being acted on, the recipient may be less inclined to rely on it, particularly where there is a prospect of financial or other loss if the action does not have the promised consequences. This is another argument for having a credible mediator between source and user of information. More generally, it supports the suggestion that governments have a role to play in establishing or supporting quality assurance schemes and regulatory frameworks which will increase the confidence with which users evaluate the information to which they are exposed.

Paying for knowledge transfer

Who should pay for knowledge transfer? Or rather, how should the cost be allocated between the various interested parties – sources of knowledge, the users of knowledge (whether individually, or through voluntary organisations, or through involuntary trade levy), commercial intermediaries, governments on behalf of the general public? Those who advocate minimum restriction on the operation of free markets would argue that those who benefit from the transfer of the knowledge should pay. In Kenya, dairies that buy milk from large numbers of small-scale producers recognise it is in their mutual interest to provide information to their suppliers that will enhance quality. They therefore invest in training and advisory materials, in the expectation of a pay-off in lower rates of contamination and spoilage. In other contexts, for example in the UK, farmers who recognise their need for information and exposure to new knowledge have organised themselves into groups to pay for services from consultants, researchers or trainers. However, given both the public good nature of most information and knowledge and the public interest in food safety and environmental quality, there remains a strong case for some investment of public resources to ensure that an optimum amount of knowledge transfer activity takes place and that producers are in a position to act in ways consistent with policy and public health.

Communication and promotion strategies

There are those who argue that a good idea needs no promotion: it will spread rapidly through the normal processes of communication and social interaction. This is about as naïve as arguing that because water flows downhill, safe drinking water will automatically find its way to everyone who needs it. In the absence of strategic interventions to remove constraints, and of proactive efforts to communicate and exchange knowledge between the various stakeholders in livestock production, existing inequalities in the distribution of knowledge will grow and science and livestock production will not learn from each other as effectively and quickly as they otherwise might. One way forward is for every research programme and project to develop an explicit communication strategy which identifies all relevant stakeholders (including other scientists, policy makers, commercial and public sector service providers, consumers, farmers) and articulates how they will engage each of them in communication throughout the life of the programme or project – not just for a one-way transmission of information, but for continuing dialogue. This can be developed as the research proceeds into a promotion strategy which identifies the specific channels which will be used to make information about the research outputs available to those who have a role to play in promoting their application in relevant contexts.

Conclusion

Communication is not a cost-free process. But it is one that can bring substantial returns and added value to the process of knowledge generation. The commercial sector knows this and demonstrates it through large budgets earmarked for advertising and other forms of promotion. It is also not an activity to which scientists can devote large proportions of their time, even if they have the inclination and expertise for it. It is incumbent on those who fund and manage research to include provision for "knowledge transfer" in their budgets and procedures, and on governments to contribute to the removal of constraints to the accessing of information by livestock producers and the creation of regulatory frameworks for assuring the quality of that information.

Acknowledgements

This paper is based in part on research funded by the LINK Sustainable Livestock Production programme under project LK0647, "Improving the targeting of knowledge and technology transfer in the livestock sector by understanding farmer attitudes and behaviour" (2001-2003), sponsored by DEFRA, The Environment Agency, MLC and MDC; and by DFID's Animal Health Programme under project R8152, "Dissemination of animal health knowledge for the development of landless dairy cattle owners in the peri-urban regions of Pondicherry, India" (2002-2004). The author acknowledges the support given by sponsors to these projects.

References

Bain, R. and Tanner, J. (1999). Healthy sheep pay the medical bills. Wambui finds out .... Illustrated Children's Story. Nairobi. Kenya Agricultural Research Institute.

Cooke, R. and Park, J. (2002). Identification of Technologies. Working paper 01/03 for LINK research project LK0647. School of Agriculture, Policy and Development, The University of Reading.

Garforth, C. (2001). Equipping the mediators: enabling extension staff in Eritrea to mediate between users and providers of agricultural information. *IAALD Quarterly Bulletin* 46 (3/4): 64-75.

Garforth, C., Angell, B., Archer, J. and Green, K. (2003). Fragmentation or creative diversity? Options in the provision of land management advisory services. *Land Use Policy* 20: 323-333.

Garforth, C. and Usher, R. (1997). Promotion and uptake pathways for research output: a review of analytical frameworks and communication channels. *Agricultural Systems* 55(2): 301-322.

Minjauw, B., Muriuki, H.G. and Romney, D. (2003). Field Schools for Kenyan dairy farmers. *LEISA* March 2003, 8-10.

MSSRF (2003). Information Village Research Project. M.S. Swaminathan Research Foundation. http://www.mssrf.org/informationvillage/infovil.html. Accessed 30th. May 2003.

Oakley, K. and Campbell, T. (2002). *On the move: a look at the social uses of mobile and wireless communications.* The Local Futures Group: London.

Rajamohan, K.G. (2003). Personal communication, K.G. Rajamohan, Scientist, M.S. Swaminathan Research Foundation, 22 March 2003.

Ramkumar, S. (2003.) Cattle rearing by the landless in the peri-urban regions of Pondicherry. Paper presented at workshop on Cattle Health Issues in the Peri-Urban Regions: Potentials of information in coping with poverty, Rajiv Ghandi College of Veterinary and Animal Sciences, Pondicherry, India, 20 – 21 March 2003. Proceedings in press.

Rehman, T., McKemey, K., Garforth, C., Huggins, R., Yates, C.M., Cooke, R.J., Tranter, R.B., Park, J.R. and Dorward, P.T. (2003.) Theory of Reasoned Action and its integration with economic reasoning in linking farmers' attitudes and adoption behaviour – an illustration from the analysis of the uptake of livestock technologies in the South West of England. Paper presented at the 14th. International Farm Management Congress, August 2003, Perth, Australia.

Riddell, I. (2001). Monitor farms and farmer discussion groups in New Zealand. SAC report. Edinburgh, SAC.

Röling, N. (1988). *Extension Science: Information systems in agricultural development*. Cambridge, Cambridge University Press. 233 pp.

Williamson, S. (2003). Pesticide provision in liberalised Africa: Out of control? Agricultural Research and Extension Network (AgREN) paper 126. Overseas Development Institute, London. January 2003, 20 pp.

20

Development of *Farmer Field School* methodology for smallholder dairy farmers in Kenya

B. Minjauw[1], H.G. Muriuki[2] and D. Romney[1]
[1]*International Livestock Research Institute (ILRI), PO Box 30709, Nairobi, Kenya*
[2]*Ministry of Agriculture and Rural Development, PO Box 30028, Nairobi, Kenya*

Introduction

Considering the escalating population growth, shrinking grazing land and increased demand for animal products, sustainable livestock production through improved management is critical to the issue of food security and poverty alleviation in most developing countries. The challenge facing the research and extension services in these countries is to increase productivity in the livestock sector while sustaining and enhancing the productive potential of the available natural resources.

In a climate of declining governmental support for conventional means of extension and the evidence of lack of success of traditional methods, the need for alternative methods for disseminating technologies is recognised (Moris, 1991; Scarborough *et al.*, 1997;Wambugu, 2001). Failure of these traditional methods can be partly attributed to the fact that they have not always focused on farmers' priority issues, or have given recommendations that were inappropriate or with no immediate tangible benefits (Heffernan and Misturelli, 2000). The underlying reason for these failings is that farmers were insufficiently involved, or not involved at all, in identifying their problems, or in selecting, testing and evaluating the possible solutions. The traditional extension system in Africa, based on a top-down approach, rarely delivering in an integrated manner the necessary information (Nyambo and Kimani, 1998). Livestock extension has focused on delivery of services such as artificial insemination, rather than delivery of information on improved management practices. It is generally acknowledged that livestock production and marketing information has been poor or non-existent, particularly in comparison with the services offered to crop producers (Barton and Reynolds, 1996). Departments of Veterinary Services have historically not undertaken extension advisory work, focusing instead on provision of emergency health care services by animal health assistants, often

Although farmers in Africa and elsewhere have expressed demand for FFS on livestock, and particularly on dairy, no curriculum has yet been developed and implementation of FFS has largely remained the domain of agronomists. Following the success of crop FFS, the FAO expressed the need to develop the methodology for livestock before implementation can take place. Experimental FFS for poultry keepers already exist in Kenya and show promising results, but it was felt that before cattle FFS could be implemented some basic research was needed. Indeed, dairy cattle represent a high investment with a slow return and, therefore, a much higher risk than crop or poultry production. Responses to health and production interventions may also be longer term, therefore requiring adaptation of the single season cycles used in FFS for crop issues (Schmidt, 1997). Furthermore, the health and nutrition issues are both quantitatively and qualitatively more complex.

Therefore, the current International Livestock Research Institute (ILRI) / Department for International Development (DFID)-Animal Health Programme (AHP)/FAO livestock FFS project is adapting and developing the methodology specifically for livestock, and particularly for smallholder dairy production, focussing on animal health and production issues. The approach is applied to developing integrated methods to control vector-borne diseases and helminth infections and to improve the efficiency of utilization of available feed resources and the management of nutrients within the crop-dairy system. Using the FFS approach, the project is developing an innovative process, through interactions between farmers, scientists and extension workers, through which farmers adapt existing technologies and try out new ideas.

The Livestock Farm Field School activities contribute to the ongoing DFID bilaterally funded Smallholder Dairy Project (SDP), led by the Ministry of Agriculture Research and Development in collaboration with the Kenya Agricultural Research Institute (KARI) and the International Livestock Research Institute (ILRI).

On-going activities in the DFID Livestock FFS project

All facilitators were trained during a 2-week training of trainers (TOT) course. The TOT was run as a learning workshop, where participants learned and at the time used the basic principles of the FFS to develop specific examples of activities, tools and techniques suitable for the smallholder dairy production systems.

Ten livestock FFS with similar characteristics (Table 20.1) and interests in dairy production were established in five different agro-ecological zones (Jaetzold and Schmidt, 1983) in Central, Rift Valley and Coastal Provinces of Kenya:

- LH3: low highland; wheat/maize-barley zone; moderately cool and semi-humid
- LH4: low highland; cattle-sheep barley zone; moderately cool and transitional
- UH2: upper highland; pyrethrum-wheat zone: cool and sub-humid
- L4: lowland; cashew nut-cassava zone; hot and transitional.
- L5: lowland; livestock-millet; hot and semi-arid.

Table 20.1. Mean characteristics of livestock FFS groups calculated from the first eight groups of farmers.

	Mean per group
Original number of male members	19
Number of female members	15.5
Total number of members	34.5
Average age of members	46
Average size of farm, ha (ac)	1.34 (3.3)
Average number of cattle	2.8
Average number of milking cows	1.5
Average % attendance	74
Average number of meetings in an 8 month period	29

Facilitators trained in FFS approaches worked with established groups to prioritize the main constraints to improved efficiency of milk production using participatory pair-wise and matrix ranking techniques. Issues highlighted for all groups were similar and, in order of priority, included: 1) Feeding strategies; 2) Forage establishment and conservation; 3) Calf rearing and mortality; 4) Diseases (tick-borne and mastitis) 5) Water management and breeding (equal priority given). Based on the results of this exercise, individual grant proposals were prepared by each group including a detailed work plan with a corresponding budget.

A grant of US$600 was deposited in an account controlled by elected members of the FFS group to cover the cost of field activities and the cost of facilitation, i.e. the transport and lunch allowances to enable the extension worker to visit. Management of this budget empowered the farmers to demand and control activities covered by the FFS and ensured that the extension services offered responded to farmers' actual priority problems and needs. The FFS groups meet normally on a weekly basis but some vary their frequency to fortnightly. The main approaches used; participatory tools applied to animal health and production issues; agro-ecological system analysis (AESA) and participatory technology development (PTD) were adapted to suit the specific needs of learning about livestock issues.

Participatory tools for problem diagnosis

As one of the first stages in the FFS, participatory methods were used to carry out in-depth problem diagnosis and develop a common understanding between the facilitator and group to determine animal production and health issues that were to be addressed. Understanding of existing farmer knowledge, local factors and the broad variation in practices between members of the group provided the basis for subsequent learning activities. Common participatory tools used in participatory rural appraisal, ethno-veterinary surveys and qualitative epidemiology (Mariner, 2001) were adapted for use in FFS, building on the participatory epidemiology approach as described by Catley *et al.* (2001, 2002a and b). Consolidation of the information across different FFS is also being used to contribute to an epidemiological study, in the area where the FFS project is operating, evaluating the use of participatory methods to develop more successful disease surveillance and control strategies.

Semi-structured interview

In participatory assessment, an interview questionnaire is not used. Instead, the facilitator prepares a checklist of important points and exercises to be covered. This allows the interview to be flexible and permits the respondents to express their thoughts in their own words within their own conceptual frameworks.

Participatory mapping

Mapping as a type of visualisation method is a popular participatory method among animal health workers and livestock keepers. Examples of maps that could be used in the dairy FFS include: resource flow maps showing movement of resources within and between farms; movement of milk produced to explore and understand marketing issues and problems; opportunities and service maps to highlight where farmers obtain services and where the deficiencies lie; social maps to show where farmers obtain information and identify social support networks. Natural resource maps combining disease incidence for a community could help to highlight community level factors that influence disease.

Mapping is a useful method for the following reasons:

- both literate and non-literate people can contribute to the construction of a map (as it is not necessary to have written text on the map)
- when large maps are constructed on the ground, many people can be involved in the process and contribute ideas. People also

correct each other, and make sure that the map is accurate
- maps can represent complex information that would be difficult to describe using text alone
- maps can act as a focus for discussion

Mapping is very useful during the ground-working stage of the FFS. The method tends to prompt much discussion and activity among farmers, and enables them to define the area under consideration. Although when copied to paper maps become useful outputs of mapping methods, it is important to note that the principal value of maps is as the focus for generating discussion and follow-up questioning.

Pair-wise ranking and matrix comparison of diseases and indicators of disease

This method is used for understanding local characterization of livestock diseases and correct identification of diseases from their local names (see Catley et al., 2001, 2002a, 2002b). The method can help to answer the question: "Are the facilitators and FFS farmers talking about the same diseases?"

Seasonal calendar

Seasonal calendars to describe cropping activities, labour use, or feed use or availability are commonly used. They can also be useful to facilitate farmers to explore issues relating to animal health. Temporal variations in disease occurrence are a common aspect of epidemiological investigation. Seasonal calendars are a practical method for understanding farmers' perceptions of seasonal variations in disease incidence or populations of ticks, biting flies or other factors (Catley *et al.*, 2002a, 2002b). Seasonal calendars can also generate new hypotheses about associations between diseases, environmental factors, and interactions with wildlife and vectors.

Agro-eco-system analysis (AESA)

The AESA exercise helps to establish, through a process of structured observation, the interaction between livestock and other biotic and abiotic factors co-existing in the field. The exercise is used to improve farmers' decision making skills by developing a system whereby regular observations of the livestock and their environment are used as a basis for identifying problems, deciding on improvements and monitoring change. At each FFS meeting farmers are divided into sub-groups who visit neighbouring farms of some of the participants. Data are collected through observation of a single animal and its interaction with the environment. Decisions on interventions to improve the health and nutrition of the animal with a view to optimizing farming objectives are made and the conclusions presented back to the whole group for critique.

The AESA takes place in four steps using a record sheet prepared by the group (see Figure 20.1).

1) General Information

   Farmers interview the owner to get general background information on the animal examined.

2) Parameters

   Farmers record all parameters that can be measured or estimated. A weigh band system is used to estimate live weight and recorded weights are compared to those of previous weeks. Milk production is recorded and reproductive status and management noted. Farmers also observe the type, quality and quantity of basal forage and supplements offered. This activity is also intended to encourage the farmers to improve their record keeping.

3) Observations

   One of the objectives of the AESA exercise is to improve the observation skills of the farmer, so that it becomes an automatic checklist that the farmer will mentally go through every time he sees his animal. Observations are performed in 3 steps: observations from a distance, close-up observation and attention to the environment.

   Step 1: Hair/coat, body condition, rumination, movement/temperament and respiration.
   Step 2: Temperature, presence of ecto- and endo-parasites, discharges, dung, urine, wounds, eye condition, mucus membrane color, lymph nodes, presence of diarrhea etc.
   Step 3: Details of the production system (zero grazing, extensive, semi-extensive), including for example infrastructure details, quality and quantity of available feed, water, shade, presence of other animal, insects, noises, etc.

4) Recommendations

   The recommendations section gives an opportunity for the farmers to suggest some solutions to improve the overall condition of the animal examined. These might include suggestions for improving infrastructure or changes that could be made to improve productivity, such as changes in forage type or amount of supplement. Farmers may also suggest how to improve the AESA through recording additional parameters or changing the way observations are made.

Aesa number
Week/date
Sub-group name

PARAMETERS
Body weight
Last weight
Weight gain:
Daily milk yield
Milk yield status:
(improving or decreasing)
Number of calves
Date of serving
Date last calving
Pregnancy status
Calving interval

Feed quality
Feed quantity
Supplement
Water quality
Water quantity

GENERAL INFORMATION
Breed
Name/tag
Sire name and breed
Dam name and breed
Date of birth and Age
Time of observation
Weather condition
Last treatment: date and drug used

OBSERVATIONS
Hair/coat
Body condition
Rumination
Movement/temperament
Respiration
Temperature
Ecto-parasites
Discharges
Dung
Urine
Wounds
Eyes condition
Mucus membrane color
Lymph nodes
Housing and shading conditions
•Presence of other animal/insects
Noises

RECOMMENDATIONS

How to improve the AESA records
•Parameter to be included
•Quality of observation

What need to be done to improve productivity?

Which treatment should be done?

Figure 20.1. AESA format developed by the facilitators during the TOT.

Livestock Participatory Technology Development (PTD)

Livestock PTDs are implemented to empower participants (both farmers and facilitators) with analytical skills to investigate cause and effect relationships of problems in farming practices. Since the main objective of the PTD is to develop farmers learning skills rather than just increase knowledge of a particular technical issue, record keeping and accurate observation are an important component. For this the AESA technique is an integral component of the PTDs and is used to record and observe the results of the experiments.

Establishment of PTD is one of the largest challenges for livestock FFS. Indeed, while it is relatively easy to design comparative studies for crop integrated pest management, the high economic value of cattle does not allow any experiment involving risk or even short-term productivity losses of the animal. Therefore, one of the objectives of the livestock FFS project, is to establish the type of PTDs that can be performed without any risk or detrimental effect and still allow farmers to experiment with new technologies. Three types of PTDs have emerged from on-going activities:

Classical PTDs

Although livestock are the focus for the FFS, a lot of activities of the livestock keeper are crop related when they are focused on addressing feed scarcity through improved forage production and grazing improvement. The PTDs that have been carried out include:

1 Establishment of alternative sources of forage. A range of forage varieties are planted using different planting methods, treatments and, or, different fertilizer regimes.
2 Preservation of forage using different techniques such as silage making and the box baler for hay are tested using material grown on the plots.

Comparison of existing farmer practices

Observation and evaluation of variation in practices used by different farmers, both within the group and neighbours, provide the opportunity to address issues where there is a high risk to animal well-being, or high costs prevent experimentation. Examples include:

1 Tick control: comparison of efficacy of different acaricides and, or different application regimes.
2 Vaccination efficacy: comparison of disease incidence in immunised and non-immunised animals.
3 Comparison of milk quality, animal health and losses due to milk spoilage in relation to the quality of the milking stall and zero grazing infrastructure and management practice.

Ex-post PTD analysis

In *ex-post* analysis, farmers compare actual experimental results with practices that were used before. Results may be quantitative, if records are available from the past or from similar situations, or qualitative, where farmer perceptions are evaluated. This also includes the "Stop and Go" method, where the treatment is stopped and re-introduced several times to show its effect using an animal as its own control. Those tested include:

1 Water availability: the amount of water available to the dairy animal is changed according to the calculated needs. Milk production on the new regime is compared with previous records of production on the old regime.
2 Genetic material: calf weights on farms where artificial insemination is used are compared with birthweights of calves from natural reproduction methods, or the expected weights.
3 Prophylactic programmes: a prophylactic programme is applied to a group of cattle and their performance is compared with previous productivity and with neighbouring herds. This could include deworming, trypanocide and, or, vaccination against prevalent diseases.

Conclusions

If scientific research is to achieve real impact on farm productivity and livelihoods, new methodologies for dissemination of information have to be developed. Participatory approaches, which facilitate farmer demand for knowledge, give the opportunity to the end users to choose, test and adapt technologies according to their needs. Through participation in FFS, farmers develop skills, which allow them to continually analyse their own situation and adapt to changing circumstances. The livestock FFS facilitates in a sustainable manner the identification and exchange of knowledge on appropriate technologies for smallholders to improve their dairy production system in environments with high disease risk and nutritional stress. The ILRI livestock FFS project is testing and adapting a participatory method to create a sustainable relation between farmers, extension officers and research institutes. These relationships are thought to be a fundamental tool for scientists to collect appropriate data and to transform developed technologies into products adapted to the end-user needs.

Not every problem can be easily dealt with by a "learning by doing" approach. Some problems, dealing with contagious diseases, for example, are not suitable or too dangerous for on-farm experimentation. Others may be too abstract to be demonstrated physically, such as the importance of epidemiological status or immunological reactions and these can be addressed in special topic sessions where issues are discussed. Since the facilitator cannot be an expert in every subject, he will help the farmer group to invite the right person to talk about the subject chosen by the farmers. This empowers the FFS group to invite other organisations, such as NGOs, national or international research institutes to conduct special topic sessions. Special topics can include livestock and non-livestock related issues, giving farmers the chance to access the information responding to their priority at a particular moment. For example, talking to the community about trypanosomiasis when the village is threatened by a cholera outbreak is unlikely to be addressing a priority issue, when advice about cholera control may be more relevant.

During the course of the project a small number of farmer groups will be affected. However, manuals and tools will be developed that can be used by other groups of people responsible for delivering knowledge so that in the long term large numbers of livestock farmers in Kenya and similar regions will benefit from the research findings.

Acknowledgements

This publication is the output from a research project (Animal Health Programme) funded by the UK Department for International Development (DFID) for the benefit of developing countries. The views expressed are not necessarily those of DFID. The financial support of FAO is acknowledged.

References

Axin, H. (1988). *Guide on alternative extension approaches*, Food and Agriculture Organisation (FAO), Rome. 148 pp.

Baltenweck I., Staal, S., Owango, M., Muriuki, H., Lukuyu, B., Gichungu, G., Kenyanjui, M., Njubi, D., Tanner, J. and Thorpe, W. (2001) *Intensification and competitiveness of smallholder dairy production systems in the Greater Nairobi Milk-Shed*. Kenya Agricultural Research Institute (KARI)/Ministry of Agriculture (MoA)/International Livestock Research Institute (ILRI) Draft Collaborative Research Project Report, 12 pp.

Barton, D. and Reynolds, L. (1996). *The generation, assembly and delivery of information on livestock production: Kenya case study*. Report for Overseas Development Administration (ODA) and World Bank, 44pp.

Catley, A., Osman, J., Mawien, C., Jones, B.A. and Leyland, T.J. (2002a). Participatory analysis of seasonal incidences of diseases of cattle, disease vectors and rainfall in southern Sudan. *Preventive Veterinary Medicine,* 53/4: 275-284.

Catley, A., Irungu, P., Simiyu, K., Dadye, J. Mwakio, W., Kiragu J. and Nyamwaro, S.O. (2002b). Participatory investigations of bovine trypanosomiasis in Tana River District, Kenya. *Medical and Veterinary Entomology*, 16:1-12.

Catley, A., Okoth, S., Osman, J., Fison,T., Njiru, Z., Mwangi, J., Jones, B.A. and Leyland,T.J. (2001).Participatory diagnosis of a chronic wasting disease in cattle in southern Sudan. *Preventive Veterinary Medicine*, 51:161-181.

Chambers, R. (1983). *Rural development: putting the last first.* Longman, London, 246 pp.

Chambers, R. (1993). *Challenging the professions: frontiers for rural development.* Intermediate Technology Publications, London, 143 pp.

Chambers R., Pacey, A. and Thrupp, L. (ed.) (1989). *Farmer first: farmer innovation and agricultural research.* Intermediate Technology Publications, London.

Garforth, C.J. (1998). Dissemination pathways for RNR research. *Socio-economic methodologies best practice guidelines.* Published by Natural Resources Institute (NRI) Catalogues Services, CAB International, Wallingford, UK, 12 pp.

Fujisika, S., Khisa, G. and Okoth, J. (2000). *FFS/IPPM contributions to sustainable livelihood in Uganda and Kenya.* Consultant's report for FAO, Global IPM Facility, 22 June 2000, 10 pp.

Heffernan, C. and Misturelli, F. (2000). *The delivery of veterinary services to the rural poor: preliminary findings from Kenya.* Report for the DFID funded Study R7359. www.livestockdevelopment.org

International Centre of Insect Physiology and Ecology (ICIPE) (1992). Adaptive research to assess the sustainability of the ICIPE tsetse super trap: an innovative tool for community based management of tsetse and trypanosomiasis in Lambwe Valley. *Annual report 1992-1993.* ICIPE, Nairobi, Kenya.

Jaetzold, R. and Schmidt, H. (1983). *Farm management handbook of Kenya. Vol II A*, Ministry of Agriculture, Kenya, 397 pp.

Jones, B.A., Deemer, B., Leyland, T.J., Mogga. W. and Stem, E. (1998). Community-based animal health services in Southern Sudan: the experience and future. Paper presented at *AITVM Conference*, Harare, Zimbabwe.

Leonard, D.K. (1985). *The supply of veterinary services. Development Discussion Paper 191.* Harvard Institute for International Development: Harvard University.

Leonard, D.K. (2000). *Africa's changing markets for health and veterinary services: the new institutional issues.* MacMillan Press Ltd., 320 pp.

Mariner, J.C. (2001). *Manual on participatory epidemiology. Methods for the collection of action-oriented epidemiological intelligence.* FAO Animal Health Manual No.10. Food and Agriculture

Organisation, Rome.

Martin, A. and Sherington, J. (1997). Participatory research methods: implementation, effectiveness and institutional context. *Agricultural Systems* 55:195-216.

Matteson, P.C., Gallagher, K.D. and Kenmore, P.E. (1992). Extension of integrated pest management for plant hoppers in Asian irrigated rice: empowering the user. In: *Ecology and management of planthopper.* Edited by Denno, R.F. and Perfect, T.J. Chapman and Hall, London.

Minjauw, B. and de Castro, J.J. (1999). Host resistance to ticks and tick-borne diseases: its role in integrated control. Chapter 7. In: *Breeding for disease resistance in farm animals, 2nd Edition* (ed. Axford, R.F.E., Bishop, S.C., Nicholas, F.W and Owen, J.B.), CAB International, Wallingford, UK, pp. 153-169.

Moris, J. (1991). *Extension alternatives in tropical Africa.* Overseas Development Institute, London, 184 pp.

Morton J., Adolph B., Ashley S. and Romney D. (2002) Conceptual, methodological and institutional issues in participatory livestock production research. *Livestock Research for Rural development* 14 (4) http://www.cipav.org.co/lrrd/lrrd14/4/mort144.htm

Nyambo, B.T. and Kimani, M. (1998). Role of information in IPM farm field schools: experience in Kenya. *Paper presented at the Integrated Pest Management Communications Workshop: Eastern/Southern Africa.* March 1-6 1998, International Centre of Insect Physiology and Ecology (ICIPE), Nairobi. Kenya.

Omore, A., Muriuki, H., Kenyanjui, M., Owango, M. and Staal, S. (1999). *The Kenyan dairy sub-sector; a rapid appraisal.* Kenya Agricultural Research Institute (KARI)/Ministry of Agriculture (MoA)/International Livestock Research Institute (ILRI) Collaborative Research Project Report.

Ooi, P.A.C. (1996). Experiences in educating rice farmers to understand biological control. *Entomophaga* 41:375-385.

Ooi, P.A.C. (1998) Beyond the Farmer Field School: IPM and empowerment in Indonesia. International Institute for Environment and Development (IIED) *Gatekeeper Series No. 78.* London. 16 pp. (Originally presented at the *International Conference of IPM - Theory and Practice, Developing Sustainable Agriculture*, Guangzhou, China June 15-20, 1998).

Rees, D., Momanyi, M., Wekundah, J., Ndungu, F., Odondi, J., Oyure, A.o., Andima, D., Kamau, M., Ndubi, J., Musembi, F., Mwaura, L. and Joldersma, R. (2000). *Agricultural knowledge and information systems in Kenya- Implications for technology dissemination and development.* Overseas Development Institute *(ODI)* Agricultural Research and Extension Network Paper 107.

Scarborough, V., Killough, S., Johnson D.A. and Farrington, J. (ed.) (1997). *Farmer-led extension: Concept and practices.* Intermediate Technology Publications, Overseas Development

Institute, London, 214 pp.

Schmidt, P. (1997). Farmer Field Schools: A participatory large scale extension approach. Record 42 of 65 - CABPESTCD 1989-1999: *Beraterinnen-News*. 1997, No. 1, 17-22.

Schreiber, C. (2002). *Sources of innovation in dairy production in Kenya*. ISNAR Briefing Paper, 8 pp.

Staal S.J., Chege L., Kenyanjui M., Kimari A., Lukuyu B., Njubi D., Owango M., Tanner J., Thorpe W. and Wambugu M. (1998). Characterisation of dairy systems supplying the Nairobi milk market : A pilot survey in Kiambu district for the identification of target producers. Kenya Agricultural Research Institute (KARI)/ Ministry of Agriculture (MoA)/International Livestock Research Institute (ILRI) Collaborative Research Project Report, 85 pp.

Sutherland, A. (1998). Participation research in natural resources. *Socio-economic methodologies best practice guidelines*. Published by Natural Resources Institute (NRI) Catalogues Services, CAB International, Wallingford, UK, 18 pp.

van de Fliert, E. (1993). *Integrated pest management farm fields school generate sustainable practices*. Paper 93-3. PhD Thesis, Wageningen Agricultural University, The Netherlands.

Wambugu, M.N. (2001). *Extension and its effect on dairy cattle nutrition and productivity in smallholder dairy cattle enterprise in Kiambu District*. MSc Thesis, University of Nairobi, Kenya.

21

Demand and supply side changes in the Santa Cruz, Bolivia, milk sector 1985 - 2002: impact on small-scale producers and poor consumers

J. Rushton[1], N. Duran[1] and S. Anderson[2]
[1]*CEVEP, Casilla 10474, La Paz, Bolivia*
[2]*Department of Agricultural Sciences, Imperial College London, Wye, Ashford TN25 5AH, UK*

Introduction

During the past three decades worldwide dairy policies have been implemented to promote the consumption of milk and milk products in urban areas and the production from rural areas close to big cities (Alderman et al., 1987). Bolivia and the city of Santa Cruz de la Sierra have been influenced by these worldwide directives and this current study examines the impact of policies on the demand and supply side of the sector, but with particular emphasis on smallholder milk producers and poor urban consumers. The current research is part of a multi-country study on the supply, demand and impact of dairy and other policies over a 10 to 15 year period (1985 to 2000) of the milk sectors of Santa Cruz, Bolivia; Nairobi, Kenya; and Kathmandu, Nepal.

Materials and methods

Bolivia has had a variety of institutions collecting information in the dairy and agricultural sector over the last two decades. Where possible data and information generated by these institutions have been employed rather than collecting primary data. The following data sources have been used:

- official government data (Ministerio de Desarrollo Humano, 1993; INE, 1997; MAGDR, 2000; INE, 2001, 2002a; 2002b; 2002c);
- government and aid sponsored research (CORDECRUZ, 1988a; 1988b; 1988c; Solano, Bermues, Ferández, Rojas and Herrero, 1998; Herrero and Rojas, 1999; Rushton, Hoyos and Sonco, 2001);
- data from the milk federation in Santa Cruz (Cadario, 2001; FEDEPLE, 2001);
- data from the Chamber of Agricultural Commerce in Santa Cruz (CAO, 2001);

- data from a private dairy farm (Ballivian, 1999); and
- a private study of dairy producers in the Santa Cruz milkshed area (Fernando Cadario, personal communication).

Where there were gaps in the secondary data sources, the following formal and informal surveys were carried out:

- a milk and milk product demand survey which collected information from 140 households in 52 districts of the city of Santa Cruz; and
- informal interviews with the management of the processing plants, supermarkets, retailers, producer retailers and dairy farmers.

Finally a search was made for information on policies, programmes and projects implemented over the last 30 years.

The policy and production information has been brought together to produce a number of policy analysis matrices (Ellis, 1992; Upton and Dixon, 1994; Shapiro and Staal, 1995) to examine the impact of policies on different parts of the milk sector.

Results

Policy and projects for the dairy sector

In the 1970s and 1980s the Bolivian government provided support for the development of the dairy sector, that concentrated on the creation of milk processing capacity and technical assistance to a small number of dairy producers in the main milk shed areas. However, an economic crisis in the mid 1980s produced major macro economic policy changes. The state dropped its role in price setting of food goods and began to withdraw from investing and managing productive sector companies. This policy shift was completed in the 1990s with the capitalization of state firms, which included the milk plants constructed with aid money during the 1980s. Since the early 1990s there has been no direct dairy policy, but some aid supported dairy projects in Santa Cruz remained until the late 1990s.

The Santa Cruz dairy sector has gone from state support, planning and management in the 1980s to a complete absence of support in the late 1990s and early 2000. Therefore, there is no longer a specific dairy policy in the country and even the leading scientists and food industrialist failed to highlight this industry as needing a direct policy (Bauer, 2000; Paz, 2000). This shift has now reached a point where milk producers are now being asked to finance support services (SENASAG) and to co-finance research (SIBTA).

Demand for milk and milk products

During the last 25 years the Department of Santa Cruz and its capital city, Santa Cruz de la Sierra, have experienced high population growth rates (4-6% per annum) and rapidly reducing poverty levels. The population has increased from 0.73 million to 1.11 million between 1992 and 2001 and the poverty levels have reduced from 47.3% to 22.4% during the same period (INE, 2002b and 2002c). The milk market in the city has grown and continues to grow due to population and economic growth doubling over the last 12 years reaching an estimated 242 million litres of milk consumed in 2002 which was equivalent to US$ 70 million. Income elasticity of demand for milk and milk products is elastic and has remained relatively unchanged over the 1990s. Price elasticity of demand for milk products is elastic in the poor groups, but inelastic in the richer income groups. Cheese is the most important product in terms of the quantity of milk consumed, whereas liquid milk and other products such as yoghurt and 'dulce de leche' are more important in terms of value.

Supply of milk and milk products

The milk producers supplying Santa Cruz can be separated into three groups: milk producers with a quota to produce milk for the main milk plant (43 million litres of milk per year, 17,000 cows in milk and 599 producers), those who produce milk for informal sales (94 million litres of milk, 51,000 cows which is approximately 5,100 producers), and those who produce milk for cheese production (105 million litres of milk per year, 85,000 cows which is approximately 8,500 producers). The latter group is the most important in terms of the quantity of milk produced and the number of cows involved in production. Producers of milk for cheese are found some distance from the city of Santa Cruz. Cows are milked seasonally and have low production levels per animal and low fertility rates. The other producers are found in areas close to the city of Santa Cruz, and the majority practise dual purpose systems with low individual milk yields and relatively low fertility rates.

The milk processing sector has expanded rapidly during the last ten years and there have also been investments in the informal producer retail sector. However, the biggest change has been the sale of the main state run milk plant, PIL Santa Cruz to the private sector. A group of producers and workers in the milk sector purchased the plant in 1996 and then sold it to La Gloria, a Peruvian subsidiary of Nestle, in 1998. The newly named plant, IPILCRUZ, dominates the formal milk sector and since it was sold to La Gloria producer milk prices have fallen both for the formal and informal producer sector (in 2001 producer prices for milk supplied to the milk plant were

US$0.21/l, the informal sector is normally a half to two thirds this price and the international milk price was US$0.27/l). The change in prices to producers appears to be having an impact on the structure of dairy production by producers supplying milk to this plant.

Analysis of official data on the import and export of milk and milk products suggests that Bolivia will become an exporter of milk and milk products in the near future. This trend and a cost analysis of milk production by producers in 2000 suggest that the producers and the processing sector are improving their competitiveness in the world market.

Prices of milk and milk products

At the beginning of the 1990s producers supplying milk to IPILCRUZ were receiving a price above that of the world market for milk. However, in 1998 producer prices for milk, fresh milk prices and cheese prices began to fall in relation to other products in the market and have not recovered since this date. The reasons for this drop are possibly: the new management of the IPILCRUZ milk plant, a reduction in demand for milk products due to an economic downturn or producers had oversupplied the market. In international terms milk producers supplying milk to IPILCRUZ now receive a price below world market prices.

PAMs and Enterprise Budgets

The formal milk sector was protected by input and output policies in the early 1990s, but is now unprotected. The policies in the early part of the decade were intended to stimulate and promote a fledgling industry, and would appear to have succeeded, as producers and the main plant seem prepared to compete on the world market. The informal sector, which is much larger than the formal sector, has not received any significant support from the state during the decade and enterprise budgets for the dairy systems producing milk for city milk traders suggest that they are having problems maintaining income levels. However, producers of milk for cheese and producer-retailers have maintained income levels per enterprise and per cow.

Conclusions

Bolivian Government actions during the late 1980s and early 1990s protected and supported a small group of dairy producers in the milk shed areas close to the city of Santa Cruz. This group of producers supply products consumed by the richer socio-economic strata of the city. However, state support for these producers has gradually been removed during the 1990s and with the capitalisation[1] of the main milk plant, these producers have seen a fall in milk prices in both 'Boliviano' and dollar terms. For the first time they are receiving milk prices below world market prices, which appears to be reducing the number of dairy farmers in the formal dairy sector and increasing herd sizes. The new owners of the milk plant have not passed the producer price reduction to the consumer, and, it appears are using part of the profits generated to expand the milk processing capacity of the plant.

The informal milk sector, which is much larger than the formal sector, has received very little state support, but has continued to develop its ability to supply fresh milk and cheese to the city of Santa Cruz. The prices for these products in relation to other goods in the market has dropped, indicating that milk supply has outstripped milk demand in the late 1990s despite the market growing due to high population growth and decreasing poverty levels. The reduction in milk and cheese prices has reduced returns per cow in the informal systems and would indicate that smallholder dairy systems are unlikely to be able to support a family to an adequate standard of living unless individual animal production and productivity in these systems can be improved. The study period has seen major policy changes. Producers supplying milk to the formal milk processing sector have been the losers in these changes as milk prices have fallen and state technical assistance support has reduced. The producers in the informal sector were not affected by previous state policy or project actions, but during the 1990s they have seen the returns per enterprise and per animal reduce. The new owners of the main milk plant have utilised their monopoly buying power to increase their profit margins. Finally, the population of the city of Santa Cruz has benefited from a greater range and quantity of milk products provided by the informal milk sector. These products are also at cheaper prices. The dairy producers in Santa Cruz have in general terms not benefited from Bolivian policy changes yet they have responded positively to increasing demand. These producers are now being requested to pay for state veterinary services and co-finance state research and extension. It would appear that their success in supplying cheaper milk and milk products is being rewarding by less support.

It is recommended that the new Bolivian government[2] needs to monitor the owners of the main milk plant to ensure they are not abusing their monopoly buying power in the formal milk sector. In terms of projects and research, there is an increasing need to support dairy farmers in searching for management methods to reduce costs and increase production and returns per cow.

---

[1] Sale of state owned and managed enterprise to the private sector.

[2] National elections took place on 30th June 2002.

Acknowledgements

This publication is an output from a research project (Livestock Production Programme) funded by the UK Department for International Development (DFID) for the benefit of developing countries. The views expressed are not necessarily those of DFID. In addition, a number of people contributed to the research: Dra. Rommy Viscarra, CEVEP; Dr. Fernando Cadario, FEDEPLE; Ing. Luis Pérez, CEVEP; Jerjes Ballivian, Las Pampitas; Dr. Fidel Villegas, LIDIVET; Ing. Rommy Peña, CIAT (Santa Cruz de La Sierra); and Dr. Mario Herrero, ILRI. Without their contributions this study would not have been possible. However, the views expressed in the article are the authors' and none of the above are culpable.

References

Alderman, H., Mergos, G., Slade, R. (1987). *Cooperatives and the commercialization of milk production in India: A literature review.* Working papers on Commercialization of Agriculture and Nutrition Number 2, International Food Policy Research Institute (IFPRI), Washington DC, USA, 90 pp.

Ballivián Cordova, J. (1999). *La Economia en Lecheria. Las Pampitas,* Casilla 3627, Santa Cruz, Bolivia, 27 pp.

Bauer, P. (2000). Política Ganadera para el Tropico Boliviano. Informe Técnico: Lineamientos de Estrategia de Desarrollo Agropecuario y Forestal. Academia Nacional de Ciencias de Bolivia, La Paz, Bolivia, pp. 69-87

Cadario, F. (2001). Censo Lechero del área integrada de Santa Cruz año 2000. Memoria del tercer simposio de Productividad Lechera. Centro Nacional de Mejoramiento Genetico de Ganado Bovino, Santa Cruz, Bolivia, pp. 15-21.

CAO (2001). Números de Nuestra Tierra 2001. CAO, Santa Cruz, Bolivia, CD ROM.

CORDECRUZ (1988a). Diagnostico de la Actividad Lechera del Departamento de Santa Cruz Tomo I Diagnostico de la Ganadería Bovina de Leche en el Departamento de Santa Cruz. Prefectura de Santa Cruz, Santa Cruz de la Sierra, Bolivia, 253 pp.

CORDECRUZ (1988b). Diagnostico de la Actividad Lechera del Departamento de Santa Cruz Tomo II Diagnostico de la Agroindustria Láctea. Prefectura de Santa Cruz, Santa Cruz de la Sierra, Bolivia., 248 pp.

CORDECRUZ (1988c). Diagnostico de la Actividad Lechera del Departamento de Santa Cruz Tomo III Diagnostico de la Comercialización de Productos Lácteos. Prefectura de Santa Cruz, Santa Cruz de la Sierra, Bolivia, 135 pp.

Ellis, F. (1992). *Agricultural policies in developing countries*. Cambridge University Press, Cambridge, UK, 357 pp.

FEDEPLE (2001). Memorias del V Congreso Ordinario de FEDEPLE Mayo 2001. FEDEPLE, Santa Cruz, Bolivia, 48 pp.

Herrero, M. and Rojas, F. (1999). Caracterizacion de sistemas de lecheria y doble proposito en Santa Cruz, Bolivia. In: *Workshop on "Los Principales Sistemas de Producción de Ganado Bovino en Bolivia"*. Edited by Rushton, J. Held on 19 and 20 August 1999 in Cochabamba, Bolivia, pp 54-63.

INE (1997), Encuesta de Presupuestos Familiares. Indices de Precios al Consumidor. CD ROM, INE, La Paz, Bolivia.

INE (2001). Encuesta de Hogares – 1999. INE, La Paz, Bolivia, 235 pp.

INE (2002a). http://www.ine.gov.bo Access date 19 March 2002.

INE (2002b). Bolivia: Distribución de la Población. Publicaciones y Datos Censo 2001. Serie 1 Resultados Nacionales Volumen 1. INE, La Paz, Bolivia, pp. 81 and pp. 32.

INE (2002c). Bolivia: Mapa de Pobreza 2001. Publicaciones y Datos Censo 2001. Serie 1 Resultados Nacionales Volumen 2. INE, La Paz, Bolivia, pp. 164.

Ministerio de Desarrollo Humano (1993). Mapa de Pobreza. Una Guía para la Acción Social. UDAPSO, INE, UPP, UDAPE La Paz, Bolivia.

MAGDR (2000). El Agro Boliviano Estadísticas Agropecuarias 1990-1999. MAGDR, La Paz, Bolivia, pp. 271.

Paz, D. (2000). Resumen Ejecutivo. Informe Técnico: Lineamientos de Estrategia de Desarrollo Agropecuario y Forestal. Academia Nacional de Ciencias de Bolivia, La Paz, Bolivia, pp. 7-18.

Rushton, J., Hoyos, G., Sonco, M. (2001). *Análisis de la Economía Pecuaria en Bolivia*. MAGDR, La Paz, Bolivia, pp. 55.

Shapiro, B.I. and Staal, S.J. (1995). The Policy Analysis Matrix applied to agricultural commodity markets. In: *Prices, products, and people. Analysing agricultural markets in developing countries*. Edited by Scott, G.J. Lynne Rienner Publishers, Boulder, Colorado, USA, pp. 73-98.

Solano, C., Bermues, A., Ferández, W., Rojas, F., Herrerp, M. (1998). Evaluación del Nivel Tecnológico en Sistemas Lecheros de Santa Cruz (Bolivia) mediante Análisis de Correspondencias Múlitples. CIAT, Santa Cruz, Bolivia, pp. 44.

Upton, M. and Dixon, A. (1994). *A guideline for policy analysts. FAO Farm Systems Management Series*. Food and Agriculture Organisation (FAO), Rome, Italy.

22

Demand and supply changes in the livestock sector and their impact on smallholders: the case of dairying in Kenya: a summary

S.J. Staal[1], M.M. Waithaka[1,2], G.A. Owour[1], and M. Herrero[1,3]
[1]*International Livestock Research Institute (ILRI), PO Box 30709, Nairobi, Kenya*
[2]*Kenya Agricultural Research Institute (KARI), PO Box 57811, Nairobi, Kenya*
[3]*Institute of Atmospheric and Environmental Sciences, The University of Edinburgh, West Mains Road, Edinburgh, EH9 3JG, UK*

Introduction

Smallholder dairy production in Kenya is one of the developing world's success stories. Kenya's improved dairy cattle herd of more than 3 million is the largest in Africa, and significantly is held mostly by smallholders. As a result, dairy production is a major part of the agricultural sector and an important source of livelihoods for at least an estimated 600,000 smallholder farm families in Kenya (Omore et *al*., 1999). Compared to its neighbours, the country has a well-developed dairy production and processing industry, and the country has broadly managed to be self-reliant in dairy products, so that except during extreme dry years, imports are negligible. About 64% of milk produced is marketed, with 36% consumed at home and fed to calves (Omore *et al*., 1999), underlining the importance of home consumption as well as production for the market. Dairy production is concentrated in the high and medium potential areas of Central Kenya, and the Rift Valley where bimodal rainfall patterns and high altitudes produce favourable production conditions. Smallholder dairy farmers account for over 50% of the total milk produced and some 80% of the total marketed milk (Staal *et al*., 2001). Like its neighbours and most other developing countries, an important feature of the Kenyan dairy industry is the importance of the informal or indigenous/traditional milk market. The informal market accounts for about 85% of the marketed milk, most of it sold directly from farmers to consumers, but also with an important role for a wide variety of small-scale traders and intermediaries.

During the 1990s policy reforms common to many countries during this period resulted in liberalization of markets for milk, dairy products and inputs, and reduced public support to services supporting the dairy sub-sector. As a consequence there has been a change in the picture of the dairy sub-sector, including relative roles of different players, market shares and market control of competitors, level and quality of services, and in the relative prices incentives. Along with larger macroeconomic trends in incomes and demographics, these have affected milk supply and demand patterns.

This summary report presents recent demand and supply trends in the dairy sub-sector in Kenya, to try to highlight consequences of liberalization and macroeconomic factors. It draws upon from studies addressing the major production and consumption zones, largely those conducted by the DFID-supported Smallholder Dairy (R&D) Project (SDP), a collaborative project between the Kenyan Ministry of Agriculture and Rural Development, the Kenya Agricultural Research Institute, and the International Livestock Research Institute. The analysis is divided into three sections; policy changes, supply side changes and demand changes.

Evolution of policies in the Kenya dairy sub-sector

Public policy in the development of the dairy sub-sector has clearly been a success in Kenya. Since independence, that policy has had two major components, a) the deliberate integration of grade dairy cattle into smallholder farming systems in the highlands (GOK, 1991), and b) creation of supporting institutions, particularly the publicly funded AI and veterinary services and a monopoly parastatal dairy processing/marketing organisation, the Kenya Cooperative Creameries (KCC). Even though at its peak in the late 1980s its share of marketed milk never rose above some 25% (the remainder was entirely handled by the informal market), the KCC acted as a "buyer of last resort" that provided market security to many smallholder farmers, and thus reduced the risk implicit in investing in the dairy enterprise. Along with strong demand for milk and the active informal market, these combined to create the smallholder dairy success story.

The period 1989 to 1992 saw the main components of the policy reform process implemented. It began with the decontrol of agricultural inputs markets (including feeds) in 1989, which have resulted in greater and more reliable input availability, and an increase in the use of purchased feeds by dairy farmers (Mbugua, 1999). In the period leading up to 1992, most public livestock services were subsidised to an equivalent of 75% (Kiriro, 2001). Withdrawal of the subsidies since then has allowed private veterinary services, typically linked to agro-vet shops, to slowly gain grounds in many dairy production areas. Some public services still function in a quasi-private manner, with farmers paying for transport and time. Reform has however resulted in patchy coverage of vet services, since private providers mainly target areas with high density of improved cattle. AI provision has however apparently declined, as private providers have not replaced reduced public services to the same degree. A key reform was the liberalisation of milk marketing in 1992, which saw the removal of the KCC's monopoly on milk sales in urban area. Since then there has been a proliferation of private milk processors, resulting in over 45 licensed new processors, although some have left the market due to intense competition, and currently some 3-5 dominate the market. Further, the liberalisation was interpreted by some to mean freedom of action for the informal market, which increased its activities particularly in urban areas. The lower prices demanded of consumers for milk in the informal market helped grow its share to some 85%, so that the formal market fell to some 15%.

Post liberalisation, the picture is now one of system still in flux, with a) patchy coverage by a strengthening private vet service, b) declining AI coverage and use in many areas with potential harmful consequences for the national herd, c) widespread and dynamic informal milk market activity, even in the heart of urban areas, and d) intense competition by formal private processors for a shrunken share of the consumer milk market.

Supply factors in the dairy sub-sector

Production factors

Farm sizes and herd sizes over the last ten years depict clearly an increase in level of intensification in dairy production, defined as a shift from mainly grazing feed systems to mainly cut and carry or stall feeding (zero grazing). This is particularly true of the Central region, where land sizes are typically less than 1.6 ha (4 acres) and herd sizes less than 3 dairy cattle (Table 22.1). This trend has been motivated by the rapid growth in rural populations. Intensification has also changed with changes in herd types, whereby dairy crosses and pure breeds have rapidly replaced local breeds, in order to produce more milk for market. Land size in only a partial constraint to this intensification, as most farmers rely significantly on forage resources gathered from off-farm, and purchased feeds.

Table 22.1. Average land and herd sizes, for households with cattle by region.

	Western	*Rift valley*	*Central*	*Coastal*
Mean land size in (ha)	1.3	2.4	1.3	8.9
Mean number of local breeds/household	3.0	1.0	0.1	2.5
Mean number of dairy crosses/household	2.0	2.0	1.0	0.3
Mean number of high grade/household	2.0	2.7	1.0	0.2

Derived from Waithaka *et al.* (2002); Staal *et al.* (2001); Staal *et al.* (1998) and Staal and Mullins (1996)

Stall-feeding system is more prevalent in the high-populated areas around Nairobi and central regions around Mount Kenya, while grazing systems are found in the Rift Valley and Coast region where land is more available (Table 22.1). Overall estimates show that use of only grazing has declined among cattle-keeping households in Kenya during the period 1990-2000 (from 39% to 23%), with semi-zero grazing (stall-feeding combined with some grazing) being the most popular method, having grown from an estimated 29% to 48% of households during the same period.

Over 80% of dairy households use acaricides to control against the prevalent tick borne diseases in most parts of Kenya. On average 30% of households have access to extension services, with state departments dominating in this (65%). This is an indication that government is still the major provider of extension services, although NGOs and the private sector play increasing roles. Over 60% of dairy households report some access to AI, although much fewer use them, with government departments providing 43% of the services. The private sector has not appreciably ventured into AI provision. Credit accessibility on the other hand has been a constraint for the majority of dairy households, with less than 20% having access to formal credit. The major formal source of agricultural finance, the Agricultural Finance Corporation, stopped issuing smallholder credit in 1998 (GOK, 1999).

Between 1985 and 1990 annual milk production doubled from 1.3 billion litres to an average of 2.5 billion litres. Since then however, during the policy reform period, output levels have stagnated and apparently fluctuated between 2.2 and 3 billion litres annually. However, these figures are aggregated from Ministry of Agriculture estimates at lower administrative levels, based on estimates of cattle populations and yields. Surveys by SDP have found that they may be significantly understating both the numbers of improved cattle nationally and milk production, and that the actual figures may be much higher. Verification of surveys in five districts suggests that cattle numbers may be double official estimates on average. Milk production has clearly increased more than officially recognized, which is substantiated by continued high observed rates of consumption.

Returns to dairy production

Gross margins per dairy households were calculated using SDP survey data. The highest revenues and gross margins were found in the Central region which is characterised by the most intensive types of production, as well as proximity to the main consumption centre of Nairobi. Capital costs are also likely to be highest, because the intensive production requires investment in zero-grazing units/stalls.

Assuming that capital costs are less than the costs of labour or feed, then these margins suggest good returns to dairy production in all zones of the country. It should be noted that family labour costs are included in this estimate, so that the margins reflect profits after family labour is paid. When compared to per *capita* average GDP for Kenya of US $345 (World Bank, 2001), these results suggest a good livelihood from dairy production. Further, they do not include the benefits farmers receive from manure used on crop land or from value of livestock as assets for financing and insurance, which would raise the levels of returns even further.

Table 22.2. Gross margin analysis.

	*Mean annual US$ per farm**		
	Western	*Rift-valley*	*Central*
Sale of Milk	187	222	453
Milk consumed at home	37	44	91
Sales of animals	22	34	81
Total revenue	247	300	625
Labour	82	72	133
Fodder & concentrate	49	66	95
AI & Vet Services	13	40	62
Other variable costs	39	37	92
Total variable cost	184	215	382
Gross margin	63	86	242

*Exchange rate of KES 79 to a dollar was used

Milk marketing

At present, direct sales to consumers, private processors, co-operative societies, and informal milk traders are the major participants in milk marketing. Informal milk traders handle an estimated 38% of all marketed milk, with raw milk accounting for 85% of the total marketed milk, while processed milk products account for only 15%, most through the Nairobi urban market. Further, the largest single share (58%) of marketed milk is sold directly to households and hotels/restaurants. As a result of the market liberalisation as well as a result of increased activity in the informal market, which offers higher prices, producer milk prices have risen rapidly after liberalisation to an estimated US$ 0.19 per litre, compared to an average of some US$ 0.10 per litre during before liberalisation. Current government policy has given little constructive attention to the informal milk market, although ILRI and partners are working with officials to find ways of improving quality and safety in indigenous markets without adversely affecting small trader livelihoods.

Policy Analysis Matrix

A policy analysis matrix (PAM) was estimated per unit of 1000 litres of milk produced (Table 22.3). The PAM method compares returns and costs under existing market conditions and policies (Private values) with those that would occur hypothetically in a neutral policy environment with no taxes, subsidies or other distortions (Social values). The results show positive private profitability, at both the farm and post farm levels. When pre and post liberalisation results are compared, private farm-level profits were seen to increase by $33 per 1000 l. It should be noted that these are regarded as "above-normal" profits, in that all costs of labour and capital have already been included, and so they suggest higher profits than available from alternative enterprises at the farm level. This shows the competitiveness of the dairy enterprise under the current technologies, input and output prices. However, social profitability after liberalisation is lower than pre-liberalisation. This may be attributed to the withdrawal of input subsidies.

At the post farm level, private profits increased even more substantially, largely as a result of deregulation of both producer and consumer prices, which allowed processors to increase their margins (and may consequently have led to the growth in the informal market, as consumers went looking for cheaper milk). Therefore, although liberalisation has brought competition Staal and Shapiro (1994), focusing on the effects of price liberalisation on Kenyan peri-urban dairy came to a conclusion that though the policy reform removed about 20-30% of negative policy effects, producer disincentives still exist. They also point-out that breaking powerful monopoly through price policy alone is difficult.

Table 22.3. PAM - Comparative analysis of the effects of pre and post liberalisation policies on the dairy industry.

		Post-Liberalisation 1992-2000		*Pre-Liberalisation 1985-1991*	
Summary Indicators	*Units*	*Farm level*	*Post-Farm level*	*Farm level*	*Post-Farm level*
Private Profitability	$/1000 l	48	216	15	61
Financial Cost Benefit Ratio		0.61	0.49	0.65	0.26
Social Profitability	$/1000 l	70	25	87	132
Domestic Resource Cost		0.53	0.89	0.33	0.13
Social Cost-Benefit Ratio		0.65	0.9	0.35	0.22
Policy Transfers	$/ 1000 l	-22	190	-72	-71
Nominal Protection Coefficient		0.9	1.79	0.35	0.59

Derived from Waithaka *et al.* (2002); Staal *et al.* (2001); Staal *et al.* (1998) and Staal and Mullins (1996)

Demand factors

One of the primary factors affecting food consumption patterns is the ability to purchase food. Dairy product and other food consumption patterns were analysed by income group using budget share and unit value equations. Kenyan households spend a large share of their budget on foods (56%), with expenditure elasticity of 0.93. Expenditure on services average 23%, much lower than food expenditure. Results show that cereals take the largest share in the households' foods budget, with dairy products being the second largest in the foods budget. This underlines the importance of milk in the Kenya diet; estimate of *per capita* annual consumption range from 80 to 125 kg, depending on location, ethnic group etc., compared with *per capita* consumption of less than 25 kg for SSA overall. Demand for milk, like other food staples, is shown to be income inelastic, underlining its role as a staple food.

The budget share of raw milk is higher than processed dairy products among low-income households. This implies that consumption of processed products increases as you move from low to high-income groups, depicting variations in purchasing powers across income groups. As shown in Table 22.4, expenditure (income) elasticity of raw milk is very low, so that higher incomes lead to only small increases in consumption, whereas elasticity of processed milk is higher. Table 22.5 shows that overall consumption of dairy products per household rises with income. Significantly, however, even very poor households below the standard poverty measure (<$1 day income) consume more than half a litre of milk per day.

Table 22.4. Parameter estimates and expenditure elasticities for dairy products.

	Expenditure elasticity
Raw milk	0.13
Pasteurised	0.50
Other milk products	1.10

Table 22.5. Average quantities of dairy products consumed per household, per month for some income groups.

Income category ($/month)	*Household milk consumption (l milk equiv /month)*
<32	20
128-256	28
>384	54

The high proportion of raw milk sales directly to consumers and through informal traders is a strong indication of income constraints to milk consumption. Price elasticity estimates found that for low income households, the price elasticity for pasteurised milk is higher (0.53) than that for raw milk (0.12), suggesting that while poor household would not reduce consumption significantly in the case of a rise in raw milk price, they would in the case of a rise in price of pasteurised milk. Increase in the formal market share is only likely to happen with broad and significant increases in household disposable income. It is useful to note that while Kenya *per capita* GDP grew at an average rate of 2.2% per year during 1990-2000, it has since stagnated. As a result demand may be slumping, which can be seen in sharp falls in farm milk prices during the flush period early in 2002.

Conclusions

Trends in the Kenya dairy sub-sector during the 1990s, a period of significant policy reform and liberalisation can be summarized as such:

- More difficult access to some services by smallholder farmers
- However, improved milk market environment due to open competition
- Good profitability of smallholder production across different settings
- Apparent stagnation in production according to official figures, but evidence from field suggest production has grown significantly
- High level of consumption of milk continues, but demand may be adversely affected by low rate of income growth
- Continued large role for informal market, and continued policy debate on how to bridge gap with regulated market

Acknowledgements

This publication is an output from a research project (Livestock Production Programme) funded by the UK Department for International Development (DFID) for the benefit of developing countries. The views expressed are not necessarily those of DFID.

References

Government of Kenya and Danish International Development Agency (1991). *Kenya dairy Master Plan.* Final Report.

Government of Kenya (1999). National Poverty Eradication Plan (1999 – 2015).

Kiriro, P. (2001). Dairy production pre and post liberalization era in Kenya and the effects of liberalization on milk production, processing and marketing. International Federation of Agricultural Producers document DP37/01.

Mbugua, P.N. (1999). *Production and use of concentrates in the smallholder dairy sub-sector in Kenya*. Smallholder Dairy (R&D) project (SDP) report. Kenya Agricultural Research Institute (KARI)/ Ministry of Agriculture (MoA)/International Livestock Research Institute (ILRI) Collaborative Dairy Research Programme, International Livestock Research Institute (ILRI), Nairobi, Kenya.

Omore, A. Muriuki, H., Kenyanjui, M., Owango, M. and Staal, S., (1999). *The Kenya dairy sub-sector; A rapid appraisal*. Smallholder Dairy (R&D) project (SDP) report. Kenya Agricultural Research Institute (KARI)/Ministry of Agriculture (MoA)/ International Livestock Research Institute (ILRI) Collaborative Dairy Research Programme, International Livestock Research Institute (ILRI), Nairobi, Kenya.

Staal, S.J., Owango, M., Muriuki, H., Kenyanjui, M., Lukuyu, B., Njoroge, L., Njubi, D., Baltenweck, I., Musembi, F., Bwana, O., Muriuki, K., Gichungu, G., Omore, A. and Thorpe, W. (2001). *Dairy Systems characterisation of Greater Nairobi milk shed*. Smallholder Dairy (R&D) project (SDP) report. Kenya Agricultural Research Institute (KARI)/Ministry of Agriculture (MoA)/ International Livestock Research Institute (ILRI) Collaborative Dairy Research Programme, International Livestock Research Institute (ILRI), Nairobi, Kenya.

Staal, S.J., Chege, L., M., Tanner, J., Thorpe, W. and Wambugu, M., (1998). *Characterisation of dairy systems supply the Nairobi milk market: A pilot survey in Kiambu District for the identification of target groups of producers*. Smallholder Dairy (R&D) project (SDP) report. Kenya Agricultural Research Institute (KARI)/Ministry of Agriculture (MoA)/International Livestock Research Institute (ILRI) Collaborative Dairy Research Programme, International Livestock Research Institute (ILRI), Nairobi, Kenya.

Staal, S.J. and Mullins. G.R. (1996). *Dairy consumption and its determinants in Coastal Kenya*. A collaborative research report between Kenya Agricultural Research Institute (KARI) and International Livestock Research Institute (ILRI), pp. 46.

Staal, S.J. and Shapiro, B.I. (1994). The effects of recent price liberalisation on Kenyan peri-urban dairy. A case study using the policy analysis matrix approach. *Food Policy*. Volume 19. Number 6.

Waithaka, M.M., Nyangaga, J.N., Staal, S.J., Wokabi, A.W., Njubi, D., Muriuki, K.G., Njoroge, L.N. and Wanjohi, P.N. (2002). *Characterization of dairy systems in the western Kenya region*. Smallholder Dairy (R&D) project (SDP) report. Kenya Agricultural Research Institute (KARI)/Ministry of Agriculture (MoA)/ International Livestock Research Institute (ILRI) Collaborative Dairy Research Programme, International Livestock Research Institute (ILRI), Nairobi, Kenya.

World Bank (2001). World development indicators. http:// devdata.worldbank.org/data-query World Bank, Washington DC.

23

Demand and supply changes in the dairy sector in Kathmandu city: impact on smallholders

P. M. Tulachan
International Centre for Integrated Mountain Development (ICIMOD), GPO Box 3226, Kathmandu, Nepal

The context

Livestock rearing in Nepal is an integral component of the farming system. It contributes about 15% to the gross domestic product (GDP) and 25% to the agriculture GDP. Milk production is therefore an important activity of the livestock sector. The majority of rural households raise one or two dairy animals for milk production. At the farm level, more than 50% of the cash income generated from livestock comes from milk and milk products.

Over the past 10-15 years there has been considerable population and economic development in Kathmandu. This has led to an increasing demand for dairy products – milk and its derivatives. Currently, more than 15 categories of milk products are sold in the city. These include a large number of imported brands of cheese, ghee, milk powder (whole and skimmed), butter and ice cream, that originate from more than 20 countries (Asia, Europe, Australia and North America).

This study examines how the increases in demand have affected smallholder dairy producers, in terms of their participation in markets. The ways that dairy policies have affected smallholders and dairy entrepreneurs are also assessed.

The objectives of the paper are to:

1) review and analyse the trends in demand and supply changes of milk and milk derivatives,
2) analyse demand for milk and milk products by different consumer types,
3) construct policy analysis matrices (PAM) to examine policy issues and constraints,
4) review and analyse impact of demand side changes on smallholders,
5) draw policy implications.

Methodology

Secondary data on milk and milk products over the past ten-year period from different sources were collected, reviewed and analysed. This was followed by a consumers' survey of 90 households; 30 households in each of three income strata – high, medium and low-income groups. Simultaneously a rapid appraisal of smallholder farmers from the Kathmandu catchment area was carried out. Another survey of milk collection centres, co-operatives and 64 sample smallholders was undertaken to identify the changes that have taken place over the past ten years in terms of milk supply, market participation, dairy herd and farm size. The data were subjected to simple analysis, descriptive statistics and correlation. To examine the impacts of government policy, PAM were constructed for two milk processing plants and two smallholder producers.

Results and discussion

Supply and demand side changes

The trend and pattern of demand over the last ten years suggests there has been a significant annual growth in the demand of milk and milk products in Kathmandu. This has not been matched by increases in supply. Increases in demand can be attributed to two factors: 1) increases in household cash income; 2) increases in population. According to the Kathmandu District Profile (1999) annual economic growth of the city is between 3.5 to 4%, while the annual population growth rate is around 4.7%. Permanent migrant settlers and seasonal migrant workers have contributed to population increases. The data show that supply changes have not been able to meet changes in demand side. There is therefore a big gap between demand and supply; the demand far exceeding the supply. As a result each year the city imports large quantities of skimmed milk powder to meet the increasing demand (Figure 23.1). This suggests a large potential for smallholders around the city to increase milk production.

Milk

Demand for milk in Kathmandu was significantly higher than supply in each year. The data show that the annual growth rate of milk supply by smallholders was 14.2%. Nevertheless, there was an annual deficit of approximately seven million litres of milk to meet the estimated demand.

Milk products

Milk products analysed were butter, ghee (churned butter), cheese,

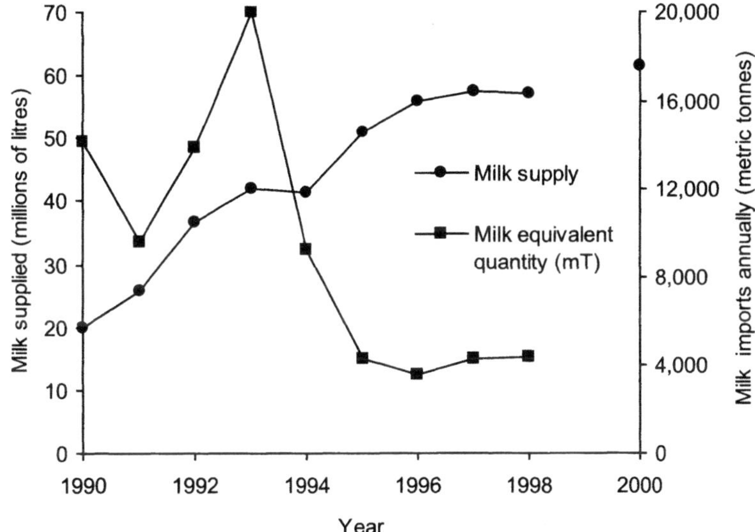

Figure 23.1. Milk supply (within Kathmandu), and milk imports into Nepal 1990-2000.

yoghurt, paneer (cottage cheese), khuwa (concentrated milk), ice cream and sweets (e.g. kalakund and ras malai). The data suggest that cheese supply increased by 26% over the ten-year period, whilst demand increased by 174%.

Initially butter supply exceeded demand, but this trend was reversed later, with demand rising massively during 1998-2000. Demand for butter rose by 150% during the ten-year period, with the demand increase following an exponential growth pattern.

Ghee demand grew exponentially until 1998 then dropped markedly. Supply increased by approximately 20% per annum, however this was irregular.

Supply of yoghurt grew by 6.2% per annum, whilst demand increased at a rate of 20 to 160% per annum. Demand and production of paneer remained relatively small in comparison to other milk products such as sweets or yoghurt. Paneer consumption remained fairly constant, with an average of 0.46 kg *per capita* consumption.

In comparison to other milk products the amount of khuwa purchased and consumed was relatively small. There was a relatively steady increase in the amount of khuwa produced, from 12.1 to 16.3 tonnes annually, between 1990 and 2000.

Demand for ice cream outstripped supply by a large amount, with demand displaying a pattern of exponential growth. Demand outstripped supply in 2000 by a factor of 59.3 or nearly 600%, indicating a large potential for developing the ice cream market in Kathmandu. Sweets made from milk were consumed on a consistent basis; supply increased from 120 to 160 tonnes per annum, whilst demand increased from 136 to 180 tonnes per annum.

Imports of dairy products

Domestic production alone was unable to meet the demand for milk and milk products. Between 80-90% of imported skimmed milk powder in Nepal was consumed in Kathmandu. The amount of domestic butter available in Kathmandu and the total amount of butter imported into Nepal showed a slowly increasing supply. However, the growth of supply was outstripped by the increases in demand for butter within Kathmandu.

Milk and milk products consumption by household income groups and monthly income

Analysis of consumers' survey data shows that consumption of milk increased considerably with increase in household income (Figure 23.2). Luxury milk products such as paneer, kalakund and ras malai also exhibited this trend; these products are consumed only by higher income groups. The amount of milk consumed displayed the most consistent increase with household income, with the amount purchased being positively related to income.

Yoghurt exhibited little variation in the amount purchased when compared to income. The amount purchased was almost constant at one litre. Changes in income levels did not appear to affect the amount of yoghurt purchased. Similarly, the amount of ghee purchased did not exhibit much variation with different income groups, suggesting a limit to the amount that can be consumed by a household.

Cheese displayed a different pattern of consumption compared to other products. The lowest income group did not consume cheese. Of the luxury products, ice cream was the most widely consumed in terms of quantity, by all income groups, with an average of 0.57 litres *per capita* being consumed weekly (Figure 23.3). The amount consumed increased with higher income groups earning above $ 401 and $ 1500.

Changes in the profiles of smallholders

Trend and pattern of dairy animals raised

Analysis of dairy herd size and composition in areas supplying milk to Kathmandu indicated an overall increase in herd size, with an increasing trend to use dairy buffaloes.

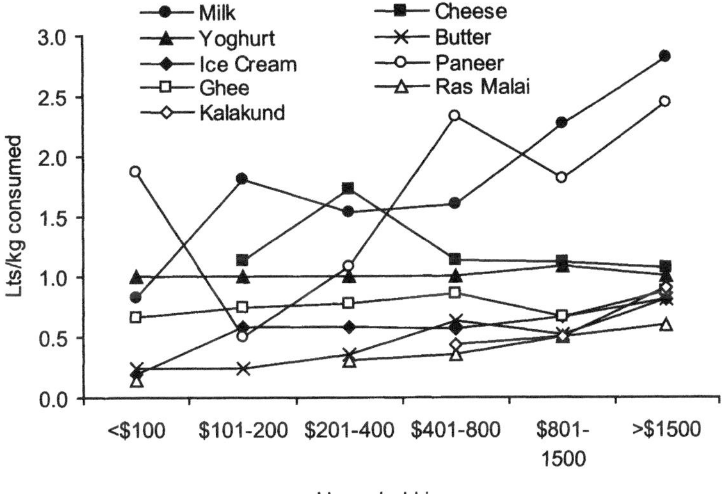

Figure 23.2. Average milk consumption by household income groups (Consumers Suvery Data).

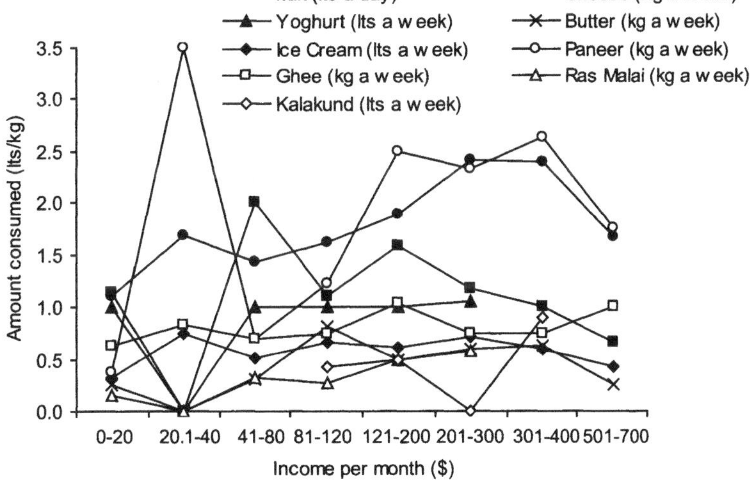

Figure 23.3. *Per capita* monthly income and dairy products consumption.

In Kavre district, dairy buffaloes increased only slightly, while the number of dairy cows remained stagnant over the past ten years. On the other hand, in Bhaktapur district, dairy buffaloes increased while dairy cows showed a decreasing trend. In Kathmandu district, dairy buffaloes increased markedly in the latter years of 1990-2000. In Lalitpur district on the other hand, dairy buffaloes decreased through the mid-nineties and there was a slight increase in dairy cows. Except for Lalitpur, the trends suggest that dairy buffaloes are playing an increasingly important role in the livestock economy.

Changes in milk output in cooperatives and collection centres

Milk collected by co-operatives and collection centres over the 1990-2000 period, showed an increasing trend (Figure 23.4). The increase in milk output was directly correlated with increases in the number smallholders supplying milk. However, milk supplied increased at a faster rate than the number of farmers supplying milk. This suggests that farmers were either increasing milk output per animal or were acquiring more animals.

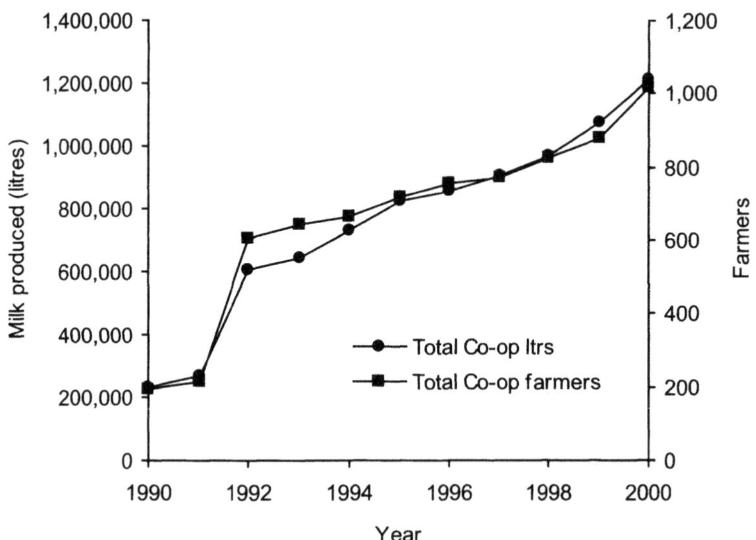

Figure 23.4. Total milk collection and number of farmers supplying the collection centres.

Smallholders supplying milk to Dairy Development Corporation (DDC)

Data from 64 sample households surveyed showed an increasing trend of milk supplied each year over the past ten years (1990-2000). However, interestingly, the capacity of milk collection from DDC (a semi-government company) went down, although the number of farmers participating increased (Figure 23.5). This encouraged smallholder farmers to sell part of their milk production to private dairy co-operatives.

Over the ten years the average area of land owned by farmers decreased yet milk production and herd sizes increased (Figure 23.6). This means that although land available per animal decreased, the actual production per animal increased.

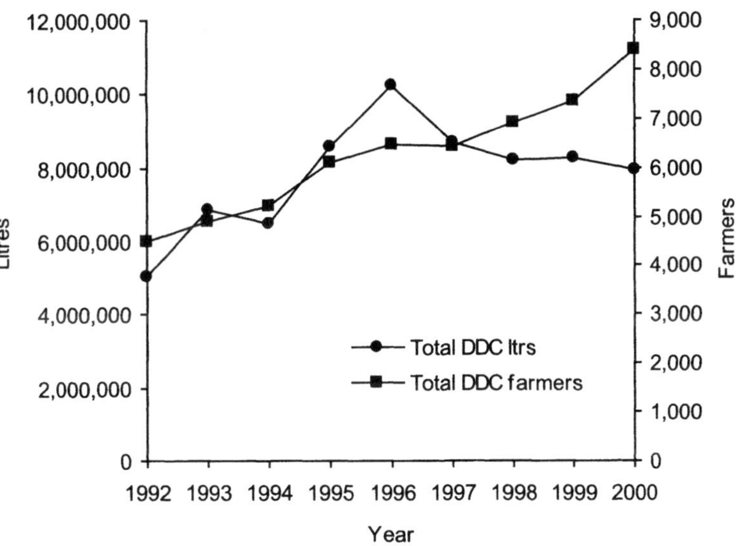

Figure 23.5. Total milk collection by DDCs and the total number of smallholder dairy farmers' participation between 1992 and 2000.

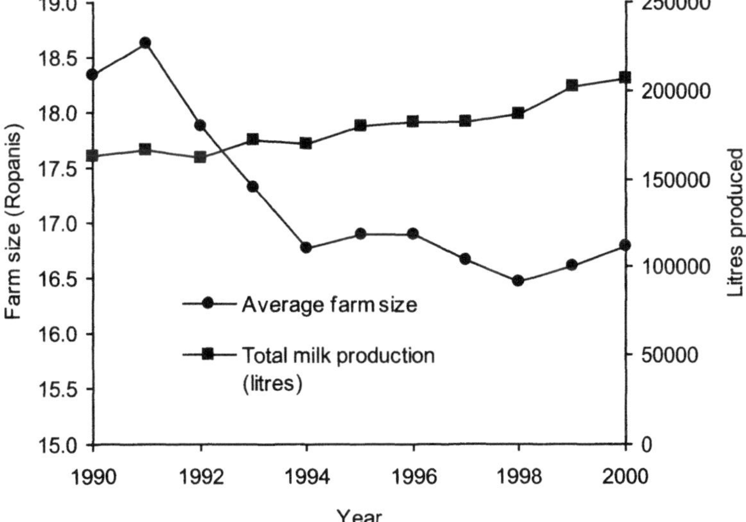

Figure 23.6. Average farm size per farmer and overall milk production amongst surveyed farmers.

Impact on smallholders

In Kathmandu and adjoining areas there are no large industries to absorb the farming labour. As a result the farming community has stagnated in the agriculture sector and mobility has been very limited. In response to increasing milk demand over the past ten years, the number of farmers and holdings selling fresh milk (delivering fresh milk to co-operatives or milk collection centres) have increased. The increase in dairy buffaloes was higher than dairy cows. There were some cases of increases in farming households due to family division following marriage of sons; each son acquiring his own land ownership from the parental property. This contributed to increases in farming households and land fragmentation leading to a decrease in household farm size (Figure 23.6). Whenever poor farmers have opportunities, they take up part-time jobs as wage labour.

Dairy policies and results of PAM

The results of PAM suggest that product diversification in terms of milk derivatives is very limited. The private sector is small and weak to bear the risk of investment in product diversification.

A small-scale milk enterprise is found more efficient because DRC (domestic resource cost) is positive which indicates room for increasing the number of small-scale dairy industries. The NPC (nominal protection coefficient), which measures the extent of policy intervention on the output side, shows that no substantial taxes are imposed on dairy products (Tables 23.1 and 23.2).

The analysis shows that the high price of milk fixed by the government has restricted development of the dairy sector. The EPC (effective protection coefficient) ratio is less than one, a sign that, despite inputs being subsidised, the overall effect of government policy is to reduce farmer incentives to be competitive in the open market.

Development of PAM for two smallholder producers (Table 23.3) suggests that the NPC, a measure of the extent of policy intervention on the output side, is 1.00 indicating that there is no substantial tax imposed on milk dairy farming. In the case of milk the EPC is 0.97 suggesting that the overall effect of government policy is to reduce farmers' incentives to be competitive in markets.

The PPC (private profitability coefficient), the ratio of domestic resources to value added both at private prices indicate that dairy farming is profitable for limited smallholders, because PPC is 0.51. The "domestic resources cost" DRC coefficient in the case of milk indicates the comparative advantage of producing it because DRC is 0.43.The private investment analysis indicates that the price of milk paid by Dairy Development Corporation (DDC) is 50% higher than its production cost. This is mainly because of the fact that milk price is being determined by DDC – a parastatal government organization. The government's policy of price fixing by DDC has discouraged the majority of smallholders to be competitive in the markets. It has also discouraged private investors to invest in the dairy sector.

Table 23.1. Milk processing enterprise: a milk plant with 300 litres of milk per day capacity – Policy Analysis Matrix.

		Tradables (US$)		Domestic factors (US$)			
		Output	*Input*	*Capital*	*Labour*	*Land / Buildings*	*Profit (US$)*
Milk	Private	2357.9	18530.7	1685.7	1435.9	884.4	1047.7
	Social	2358.5	16846.2	1169.6	1561.2	786.8	2891.0
	Difference	-0.6	1684.5	-516.1	-125.3	97.6	-1843.3
Yoghurt	Private	861.0	652.0	61.4	52.4	32.3	63.0
	Social	1066.9	592.7	67.9	70.6	35.6	30.0
	Difference	-205.9	59.3	-6.44	-18.2	-3.3	32.9
Ghee	Private	169.2	126.9	12.1	10.25	6.3	13.7
	Social	184.6	115.4	11.8	12.25	6.2	39.0
	Difference	-15.4	11.5	0.3	-2	0.1	-25.3
Paneer	Private	3743.6	2470.8	267.5	227.9	140.4	637.1
	Social	3743.6	2526.9	237.9	247.7	124.8	606.1
	Difference	0	-56.2	29.6	-19.8	15.6	31.0
Ice cream	Private	701.9	56.2	29.6	42.8	26.4	466.6
	Social	818.5	105.3	52.2	54.3	27.4	579.4
	Difference	-116.6	-49.1	-22.6	-11.5	-1.0	-112.8
Protection (Trade) Measures:							
	Nominal (NPC[1])	1.00	0.81	0.92	1.00	0.86	
	Effective (EPC[2])	0.75	0.44	0.61	1.05	0.82	
Profit (Efficiency) Measures:							
	Private (PPC[3])	0.79	0.70	0.68	0.50	0.20	
	Social (DRC[4])	0.18	0.37	0.44	0.50	0.19	

[1]NPC (nominal protection coefficient); [2]EPC (effective protection coefficient); [3]PPC (private profitability coefficient); [4]DRC (domestic resource cost)
The above five sets of coefficients and cost are estimated/computed for milk, yoghurt, ghee, paneer and ice-cream, respectively

Table 23.2. Milk processing enterprise: a milk plant with 1000 litres of milk per day capacity – Policy Analysis Matrix.

		Tradables (US$)		Domestic factors (US$)			
		Output	*Input*	*Capital*	*Labour*	*Land / Buildings*	*Profit (US$)*
Milk	Private	61769.2	29557.5	34934.1	2443.7	2256.2	7422.7
	Social	61769.2	26868.5	31440.6	2688.1	2030.9	1259.0
	Difference	0	2689.0	3493.5	-244.4	225.3	6163.7
Yoghurt	Private	16846.2	12134.2	14341.3	100.2	926.4	6163.7
	Social	17969.2	11030.2	165.5	1103.5	833.8	7905.5
	Difference	-1123.0	1104.0	14175.8	-1003.3	92.6	-1741.8
Ghee	Private	17501.3	1.9	10388.3	726.7	673.6	3079.3
	Social	19092.3	799.1	9349.5	799.3	603.9	348.5
	Difference	-1591.0	-797.2	1038.8	-72.6	69.7	2730.8
Paneer	Private	14974.4	7156.0	8457.7	591.6	546.3	1777.4
	Social	14974.4	6505.0	7611.9	650.8	491.7	285.1
	Difference	0	651.0	845.8	-59.2	54.6	1492.3
Ice cream	Private	31586.5	20145.7	23810.3	59.2	1538.1	1492.3
	Social	36505.6	18313.0	21429.3	1832.1	1384.3	3221.0
	Difference	-4919.1	1832.7	2381.0	-1772.9	153.8	-1728.7
Protection (Trade) Measures:							
	Nominal (NPC[1])	1.00	0.94	0.92	1.00	0.09	
	Effective (EPC[2])	0.92	0.68	0.78	0.92	0.03	
Profit (Efficiency) Measures:							
	Private (PPC[3])	1.23	3.45	1.35	1.23	2.36	
	Social (DRC[4])	1.04	2.14	0.97	1.03	0.07	

[1]NPC (nominal protection coefficient); [2]EPC (effective protection coefficient); [3]PPC (private profitability coefficient); [4]DRC (domestic resource cost)
The above coefficients and cost are estimated/computed for milk, yoghurt, ghee, paneer and ice-cream, respectively.

Table 23.3. Milk producer in an integrated crop-livestock mountain farm – Policy Analysis Matrix.

		Tradables (US$)		Domestic factors (US$)			
		Output	*Input*	*Capital*	*Labour*	*Land / Buildings*	*Profit (US$)*
Milk	Private	726.9	179.1	175.4	80	25.6	266.7
	Social	726.9	161.2	175.4	40	23.1	327.2
	Difference	0	17.9	0	40	2.5	-60.5
Protection (Trade) Measures:							
Nominal (NPC[1])		1.00					
Effective (EPC[2])		0.97					
Profit (Efficiency) Measures							
Private (PPC[3])		0.51					
Social (DRC[4])		0.43					

[1]NPC (nominal protection coefficient)
[2]EPC (effective protection coefficient)
[3]PPC (private profitability coefficient)
[4]DRC (domestic resource cost)

Conclusion and policy implications

The study suggests that there has been considerable growth in demand for milk and its derivatives over the past ten years in Kathmandu city due to increases in both population and household cash income. In response to the growing demand there has been an increasing supply of milk in particular and other dairy products each year through greater participation of smallholders and the private sector in the markets. The demand far exceeds the supply. The supply from smallholder producers and private entrepreneurs in the city and from adjoining areas has not been able to meet the demand. As a result each year the city imports a large quantity of dairy products to satisfy the growing demand. This shows a great opportunity for smallholders to produce more milk and increase farm income. However, the present incapacity of the private sector in product diversification and government controlled pricing policy seem to be constraining the rapid growth of the dairy sector. The results of PAM show that the overall effects of the government policy are not conducive to smallholders increasing milk supply and to be competitive in the markets.

Acknowledgements

This publication is an output from a research project (Livestock Production Programme) funded by the UK Department for International Development (DFID) for the benefit of developing countries. The views expressed are not necessarily those of DFID.

References

Agriculture Projects Services Centre (1986). *A study report on milk pricing policy and establishment of price review mechanisms for Dairy Development Corporation*, Kathmandu, Nepal.

Agricultural Statistics Division, Department of Food and Agricultural Marketing Services (1989/90 to 1999/2000). *Agriculture statistics of Nepal,* Lalitpur, Nepal.

Gotsch, C. (1989). *Agricultural policy analysis on electronic spreadsheets: Commodity and farming systems models.* Farm Management and Production Economics Service, Agricultural Services Division, Food and Agriculture Organisation (FAO), Rome, Italy.

New Era (1990). *Consumption of milk and milk products (A study of hotels/restaurants, tea shops, schools/hospitals and supermarkets/coldstores).* New Era, Kathmandu, Nepal.

Department of Agriculture Development, Livestock Development Division, PLBP (1993). *Livestock activity recording farm economics.*

FAO (1994). *Farm systems management series: Farm and community information used for agricultural programmes and policies,* Food and Agriculture Organisation (FAO), Rome, Italy.

FAO (1994). *Farm systems management series: Methods of micro-level analysis for agricultural programmes and policies: A guideline for policy analysis,* Food and Agriculture Organisation (FAO), Rome, Italy.

Gautam, K.R (1989). *Implementation and administration of agricultural price policy in Nepal.* Selected country studies on implementation and administration of agricultural price policy in Asia. Volume II, Food and Agriculture Organisation (FAO), Rome, Italy.

Monke, E. and Pearson, S. (undated). *Evaluation policy choices in developing countries: The Policy Analysis Matrix.*

NDDB (2001). *National milk marketing and strategy study: Final report.* Harihar Bhawan, Lalitpur.

NDDB (2001). *National milk marketing and strategy study: Annexes.* Harihar Bhawan, Lalitpur.

National Planning Commission (1995). *Agriculture Perspective Plan.* Kathmandu, Nepal

Shrestha, H.R (1999). *A study on smallholder dairy farming systems at milk shed area of Kathmandu Milk Supply Scheme.* International Centre for Integrated Mountain Development (ICIMOD), Kathmandu, Nepal

Shrestha, H.R, Shrestha, R.P. and Pradhan, S.M. (1992). *A comparative study on the performance of crossbreds (F1) dairy cattle.* Proceedings of the 6th Austro-Asia Animal Science Congress. Volume III, Bangkok.

Tulachan, P.M.and Neupane, A. (1999). *Livestock in mixed farming systems of the Hindu Kush-Himalayas: Trends and sustainability,* International Centre for Integrated Mountain Development (ICIMOD), Kathmandu, Nepal/Food and Agriculture Organisation (FAO), Rome, Italy.

Tulachan, P.M, Saleem, M, Jabbar, M. (2002). *Smallholder dairy in mixed mountain farming systems of the Hindu Kush-Himalaya,* International Centre for Integrated Mountain Development (ICIMOD), Kathmandu, Nepal.

Upton M. and Dixon, J.M. (1994). *Methods of micro-level analysis for agricultural programmes and policies: a guideline for policy analysts.* Food and Agriculture Organisation (FAO), Rome, Italy.

A comparative analysis of milk sectors - Nairobi, Kathmandu, Santa Cruz: Opportunities for pro-poor policies

S. Anderson[1], J. Rushton[2], P. M. Tulachan[3], S. Staal[4] and M. Herrero[4,5]

[1]*Department of Agricultural Sciences, Imperial College London, Wye, Ashford TN25 5AH, UK*

[2]*CEVEP, Casilla 10474, La Paz, Bolivia*

[3]*International Centre for Integrated Mountain Development (ICIMOD), GPO Box 3226, Kathmandu, Nepal*

[4]*International Livestock Research Institute (ILRI), PO Box 30709, Nairobi, Kenya*

[5]*Institute for Atmospheric and Environmental Sciences, University of Edinburgh, Edinburgh, EH9 3JG, UK*

Introduction

The paper presents the main findings of a set of comparative milk sector surveys carried out in Kathmandu (Nepal), Nairobi (Kenya) and Santa Cruz (Bolivia). The surveys assessed changes in demand, supply and marketing of milk and milk products over a ten to fifteen year period to 2001. The paper considers demand for milk and milk products, changes in milk supply and trading pathways, the policy frameworks, changes in demand and market share, and growth in the sector and poverty reduction. The policy implications for pro-poor development of the milk sector are discussed.

Demand

Demand for milk and milk products in the urban areas surveyed is increasing due to population increase, and where economic growth is greater than population growth, due to an increase in income. The surveys showed that demand is differentiated both in terms of quantity demanded and types of products consumed across income groups:

- Poor households consume less liquid milk and less processed products in terms of monetary value and quantity than do richer households. However, the proportion of disposable income spent on milk by poor households is often greater than by rich households;

- Raw milk budget share decreases with increased income (71%, 38% and 28% for low-, middle- and high-income groups respectively in Nairobi), and demand for processed (and in some cases imported) milk products increases markedly with household income;
- Higher income groups are most sensitive to price increases in raw milk prompting a substitution with other liquid milk products. Whereas, low-income households reallocate expenditure from other items to raw milk (Nairobi) and fresh cheese (Santa Cruz);
- Price elasticity of demand for higher value processed milk products is elastic for the poor groups, but inelastic in the richer income groups. Poor households select the cheapest milk products to maintain consumption; in Santa Cruz this was fresh cheese, Nairobi raw milk, Kathmandu raw and fermented milk.

It is noted, therefore, that developments within the milk sector that promote the production of processed milk products are not addressing the needs of poor urban consumers. Where such developments have been combined with restrictive marketing of milk and milk products there are liable to be negative effects on the poor consumers.

Supply-side changes

Changes on the supply-side are due to farmers entering and leaving the sector, changing their production strategies and adopting technology. Farmer numbers in the specialist dairy sector in Santa Cruz are static or falling, whereas the numbers of farmers in the dual purpose (milk and beef) sector are increasing. In part this is due to the intensive dairy system in Santa Cruz having higher costs and lower gross margin per animal unit and per hectare despite higher per animal output, than the dual-purpose system.

Past dairy development has focused on capital intensive strategies. However, where capital is relatively expensive with regard to labour it makes little sense to push dairy development projects that replace labour with capital. Under such circumstances capital intensive systems can only compete with existing labour intensive systems if they are subsidised.

The numbers of farmers selling fresh milk for consumption in Kathmandu (delivering fresh milk to co-operatives or milk collection centres) have increased markedly over the last decade. The composition of the Nepalese dairy herd has changed through a significant increase in the buffalo proportion, this whilst official programmes in the region have been pushing crossbred cattle. Official census data from central Nepal show that buffaloes produce twice the amount of milk per head as cattle (approximately average 950 l vs. 450 l). Dairy animals per producer have increased slightly also. In Kenya, technological innovation by smallholder cattle producers has occurred. Part of this innovation is due to the adoption of technology generated by research. Farm and herd sizes show an increase in intensification of milk production (output per animal, and output per hectare) particularly after liberalisation in 1992. Central and Kiambu districts have high concentrations of intensified dairy production, with land sizes of less than 1.6 ha (4 acres) and herd sizes less than 3 dairy cows. Intensification has also meant changes in breed types; dairy crosses and pure breeds have largely replaced local breeds. Credit accessibility has been a major drawback for the majority of dairy households, with less than 20% having access. The high rate of hired labour use by dairy households highlights the important contribution by dairying in income generation and employment creation.

Thus, smallholders have responded positively to increasing urban demand by either increasing productivity (Bolivia, Nepal and Kenya) and/ or accepting lower profit margins (Bolivia and Kenya).

Pathways for milk trading

Milk is traded through both formal and informal pathways. An informal pathway is defined here as where some or all of the actors involved in milk production, collection, processing and distribution operate outwith the sector that is regulated and where taxes and subsidies apply. Others have referred to the informal sector being 'unorganised' or 'raw and traditionally processed milk markets', neither of these definitions are precise enough to allow a discussion of policy implications.

Table 24.1 shows the main pathways for the trading of milk encountered in the surveys.

Table 24.1. Milk trading pathways.

	Total output (billion litres)	*Traded - formal markets*	*Traded - informal markets*	*Traded by producer-retailers*	*Used by the producer household*
		% of total output			
Nepal	1.1	10	2	88	
Kenya	1.7	12	17	35	36[A]
Santa Cruz	0.27	17	38	35	10[B]
Average		13	19	35	23

[A] Including residual milk suckled by the calf following milking.
[B] Not including the residual milk suckled by the calf following milking.

The importance of the informal and producer-retailer marketing of dairy products can be appreciated from Table 24.1. The trading of nationally produced dairy products through the formal sector accounts for less than a sixth of milk output. In the cases of Kenya and Santa Cruz, Bolivia, the trend has been for the proportion of milk traded through formal channels to decline over the last ten years. In Nepal, although data are less precise, the proportion traded formally has slightly increased.

The existence of the informal market is due to the unwillingness of many poor consumers to pay the extra cost of processed and packaged milk products (home processing is preferred), and to producers taking advantage of selling milk through a pathway that has lower costs and hence lower market margins between producer and consumer.

The informal sector in all countries appears well equipped to satisfy increasing demand for less processed and lower priced products. The larger-scale producers and processors are best placed to take advantage of demand increases for processed products where incomes rise

Policy framework

In Nepal, milk products have a minimal export tax of 0.5%. Milk and yoghurt imports are not taxed. All other milk products have a 10% import tax. Public investment in the Dairy Development Corporation (owned by HM Govt.) has stopped, but privatisation has not been implemented due to political pressures.

In Kenya, following the deregulation of agricultural input markets in 1989, maize marketing in March 1993, and the removal of value import duties and value added tax on agricultural inputs in 1993, there has been a growth in the private sector involvement in processing and distribution in the major milk producing regions. Deregulation of input prices and markets has led to an improved availability of feeds and a more efficient private sector feed supply at higher prices. Reform of service provision to agriculture in 1992 has resulted in patchy coverage of veterinary services, since private providers mainly target areas with high density of improved cattle. Liberalisation of the milk market has led to consumer prices increasing much more than producer revenues.

In the early 1980s a recession in the world mineral markets put pressure on Bolivian government revenues and led to a liberalisation of the national food markets. The capitalisation of industries (i.e. form of privatisation of state owned enterprises), including the milk plants in the major cities, was implemented during the mid-1990s and has placed the major state enterprise in the hands of the private sector. Since the capitalisation of the milk plant in Santa Cruz, the private sector has used its monopoly purchasing power in the raw milk market to reduce prices paid to farmers for milk.

♦

The three countries studied have experienced differing degrees of market liberalisation and the privatisation of milk plants over the study period. These policies seem to have (inadvertently) strengthened the informal sector, increasing its market share and possibly increasing the level of consumption of milk and milk products. However, integrated (consumer and producer) pro-poor policies for milk sector were not encountered in the regions surveyed.

The evidence from the surveys reported here question the belief, by governments that milk consumption and milk production by smallholders could be encouraged by milk sector policies supporting milk plant investment and dairy cooperatives, is appropriate in all situations.

Table 24.2 summarises some of the findings of the policy analysis matrix calculations made on the information from the surveys. Economic efficiency (comparative advantage) of the milk sectors both at the farm and post farm-gate level is assessed, as is the protection afforded to domestic milk producers and processors by the different governments' policies.

Table 24.2. Policy analysis results for farm and post farmgate.

Case	*Domestic cost ratio (DCR)*[A]		*Nominal protection co-efficient (NPC)*[B]	
	Farm	*Post farmgate*	*Farm*	*Post farmgate*
Kenya	0.53	0.89	0.90	1.79
Nepal	0.43	1.02	1.00	1.00
Santa Cruz	0.89	1.03	0.98	0.95

[A] DCR is an indicator of the economic efficiency of the sector or the comparative advantage and is calculated as a ratio of primary domestic factors at economic prices (i.e. land, labour and capital valued at economic opportunity cost) to value added in economic prices (world prices).
[B] NPC is an indicator of the degree of protection afforded by government policy to the sector and is a ratio of the domestic and world input and outputs prices.

Changes in demand and in market share

The surveys provide information on the population and economic growth of the urban centres[1]. From these data and milk consumption

[1] Kathmandu: annual population growth rate 4.7%, annual economic growth 3.5%; Nairobi: annual population growth rate 4.8%, growth in Kenyan GDP <2%, -0.3% in 2001; Santa Cruz: 0.73 to 1.11m 1992-2001, annual economic growth > 5%

data, preliminary estimates were made of the likely growth in demand for milk and milk products over the 5 year period 2002 to 2007. Using the information obtained on the ratio of formal to informal milk trading patterns estimates were made of the growth likely in each market. Table 24.3 summarises the findings.

Table 24.3. Preliminary estimates of growth in demand for milk and milk products in formal and informal markets.

	Population change 2002 to 2007	*Change due to per capita change*[B] *(%)*	*Change due to population increase (%)*	*Aggregate change (%)*	*Market*[C]	*Change in market share 2002-2007 (2002 = 100)*
Kathmandu	1.1 to 1.37m[A]	0	25.5	25.5	Informal	119.0
					Formal	96.3
Nairobi	2.3 to 2.91m	-11.5	26.5	15.0	Informal	113.6
					Formal	96.0
Santa Cruz	1.11 to 1.41m	5.8	27.0	32.8	Informal	130.3
					Formal	99.3

[A] Considered to be a conservative estimate of population increase
[B] Where economic growth < population growth then the assumption was that 1 per cent per year of middle income households move into poor strata; where economic growth > population growth then 1 per cent per year of poor income households move into medium strata; where economic and population growth are equal no change expected in relative proportions of the strata.
[C] From the evidence available it was assumed that the informal market would gain 1% per year from the formal market.

In Kathmandu the estimates for growth in demand are due to population growth. It is predicted that by 2007 the informal market for milk will increase by close to 20% and that the formal market will shrink slightly. In the case of Nairobi, despite the fall in demand due to the decline in economic growth, aggregate demand will rise by 15% due to population growth. Most of this additional demand is predicted to be satisfied by the informal milk market. For Santa Cruz, it is estimated that economic and population growth will lead to demand for milk increasing by a third of the 2002 total and the bulk of this increase will be met by the informal market, with the formal milk market maintaining its overall market share.

Growth in the milk sector and poverty reduction - policy lessons

Table 24.4 summarises some of the policy lessons from the surveys and other sources.

Different reports have identified the increasing demand for livestock products, including milk, as a potential driving force for reducing

Table 24.4. A summary of issues, costs, benefits and policy lessons from the milk sector surveys.

Issue	Costs	Benefits	Policy lessons
Informal Processing & Marketing	Regulation and taxing impossible;	Agile sector able to take advantage of demand from poor consumers; Low costs and prices	Regulation unlikely to achieve pro-poor benefits
	Little or no internalisation of costs caused by poor milk hygiene and/ or quality; No insurance or indemnity	Greater variety of regional and traditional products	Risk assessment necessary; Awareness of risks required by informal suppliers & consumers
	Growth limited by risk of being undermined by state; Wages not controlled and may be inequitable;	Growth & response to demand changes as state imposed costs do not apply; Labour rather than capital intensive;	Recognition of informal sector and provision of enhanced trade-offs for entering formal sector – lower costs of regulation, credit availability etc.
Formal Market	Major costs to entry; and in some cases high administration costs to maintain formal status Products of higher costs and less accessible to poor consumers Rent seeking by public sector officials Higher costs and hence prices	Regulation and taxation possible	Need to reduce the transaction costs of the formal sector

Table 24.4. Contd.

Issue	*Costs*	*Benefits*	*Policy lessons*
Privatisation	Processes have tended to remain incomplete and ineffective – capital resources often remaining in the hands of a few	Less costly to the state and expectation of a diversification of services	Carry through coherent policies
	Implementation failure	Various possible strategies including – marketing, services and producer coops	Better training of officials in processes
	Confidence for investment requires time to build up	Private sector often able to access market information better than public sector	Make policy framework coherent and stable
Milk Producer Organizations	Ineffective without market opportunities Cost of products increased by processing	Regulations easier to impose with organised groups Value adding through processing	Most appropriate where large processors, strict regulations and weak demand for liquid milk exists
	Public support often needed for start up Investment in management expertise often required		Least appropriate where there is poor infrastructure, long distances and high demand for liquid milk
Regulations on milk hygiene etc.	Zoonotic hazards particularly in immunity challenged population Hard regulations force market agents into informal market	Milk products can be sold at higher prices and exported	Build hygiene culture through investment in awareness Self regulation an unexplored option

Table 24.4. Contd.

Issue	*Costs*	*Benefits*	*Policy lessons*
Importation of milk	Informal processing sector not able to benefit from imports		Livelihood trade-offs need to be considered as imports unlikely to benefit poor producers/ retailers, and often used for higher priced products
	Protection of the milk sector expensive	Milk powder might be used to cover deficit supply in formal sector	

poverty. It has been suggested that urbanisation accompanied by economic growth has and will lead to rises in real wages, and that in turn this will enable an increase in the demand for livestock products. The supply response will involve a tightening of labour markets and further increases in real wages. This 'virtuous' cycle is enhanced by productivity gains in the milk sector where a comparative advantage exists.

However, leakages from the cycle can come about if:

- the increased employment and wage income lead to limited increases in demand for local milk products but increased demand for processed and/ or imported milk products. From the results of the survey this leakage is likely to be greatest where incomes only improve for the middle and upper strata;
- the local milk supply is inelastic and unable to respond to demand increases then benefits will accrue elsewhere. Restrictions imposed by supply and marketing regulations are liable to make supply less elastic. Notable in this regard has been the way India dairy policy has been laissez-faire towards the informal sector which now is one of the largest anywhere;
- local supply of milk is unable to compete on price with imported milk and milk products;
- the supply response is capital intensive then growth in milk production leads to only limited job creation. For example, public investment in milk plants (directly or through subsidies) will result in unfair competition to the informal sector. The milk supply may be adequate for the better off urban consumers, but few jobs are created and poorer consumers are left with little option but to buy expensive milk products;
- the benefits of the increased demand cause an inequitable distribution of assets and incomes. This is more likely if a dairy cooperative model is used. Wealthier farmers benefit from greater organization, processing power and wealthier consumers benefit from locally processed products which are probably cheaper than imports.

Expansion and productivity gains in local milk production may reduce leakage since local demand for livestock products increases rapidly with rising incomes.

Potential contributions of smallholder milk production to poverty reduction may be through:

- growth in traded milk production (due to productivity or price advantages) sucking in labour. This is dependent on the smallholder dairy producers and the informal processing sector having direct

or indirect access to markets. Evidence from the surveys suggests that the informal processing and marketing sector has facilitated market access for SHD producers;
- growth in productivity of milk sector, with raising of real wages;
- higher rural wages leading to demand for milk products, creating further jobs in response to growth elsewhere.

All of the above require local labour intensive livestock production to 'out compete' imports and capital intensive local livestock production. To be competitive small-scale milk production relies on relatively low cost labour intensive forms of production. The transition into capital intensive production will depend on the labour/ capital cost ratio. This can be monitored for. Policy support for labour or capital intensive development of the supply side requires the identification of when other policies such as credit availability are introduced.

Evidence has been presented (FAO ILRI, 2001) that wage levels in small-scale milk processing and marketing are low. In addition, employment in the informal processing and marketing sector is prone to seasonal fluctuations in milk supply and demand. However, the numbers of people employed are higher than might be anticipated (between 1 and 10 people employed per 100 litres of milk traded (data from Kenya, Ghana and Bangladesh), and it is positively correlated to the amount of processing involved. In Bolivia it has been estimated that three jobs are directly created for every 100 litres produced, one directly related to the milk production and two indirectly related (Vaca Diez, 2002).

Milk powder imports might be used to cover periods of deficit supply in the formal sector. However, favouring the formal sector in this way may offset the livelihood benefits to poor producers (making them less competitive) and processors (less able to utilise milk powder) in the informal sector. If reconstituted products are similarly priced to other processed from the formal sector they will be inaccessible to poor consumers. Evidence from the surveys of a displacement of domestic products by imports was seen in the Bolivian and Nepalese cases.

For small-scale entrepreneurs to have the confidence and ability to develop milk businesses following a liberalisation and privatisation process takes a long time. Policies should therefore seek to instil confidence and build business acumen.

Growth in the milk demand can occur in one or both of the informal sector or the formal sectors. The surveys show that the informal sector has responded best to the increasing demand in Kenya and Bolivia, and in Nepal both sectors have grown.

The evidence available suggests that under certain conditions (high local demand for liquid milk or products requiring little processing, poor infrastructure and long distances) the informal will out-compete the formal sector. For example, in Santa Cruz the informal sector competes well where the main demand is for fresh cheese even though the production is some distance from the urban centres. Added to this, employment opportunities in the informal sector are often greater where processing and marketing are more labour intensive. Given these considerations the regulation of the informal sector would be expensive and is unlikely to be successful.

If governments want informal producer/ retailers and processors to move into the formal sector the transition process needs to be subsidised and low cost. Once in the formal sector the costs of maintaining this status need to be kept to a bare minimum. If regulation of the sector is for tax collection the process may be counterproductive as prices to consumers may increase restricting consumption by poor consumers. Collection of tax from small producers is often more costly than the value of the taxes generated. If the tax revenue is to contribute to public health issues, then proper animal health surveillance to assess zoonotic risks is required, as are estimates of the costs of controlling the sector and the benefits from control.

Further evidence is required on the trade-offs involved in a policy framework that recognises the contribution of the informal sector to economic growth (supply response at accessible prices) and poverty reduction (through employment creation) via milk production and trading. Preliminary evidence from the three surveys suggests that the costs of zoonotic hazards from milk are reduced by home processing, which is common. In addition, research in Kenya shows that the hygiene of milk in the informal sector compares with that of the formal sector.

Acknowledgements

This publication is an output from a research project funded by the UK Department for International Development (DFID) through the PASS Livelihoods Programme for the benefit of developing countries. The views expressed are not necessarily those of DFID.

References

FAO (2001). Report of the FAO E-mail conference on small-scale milk collection and processing in developing countries. 29 May-28 July 2000, Food and Agriculture Organisation (FAO), Rome, Italy.

FAO ILRI (2001). Employment generating through small-scale dairy marketing and processing: Experiences from Kenya, Bangladesh and Ghana. Rome/ Nairobi, December 2001, Food and Agricuture Organisation (FAO) and International Livestock Research Institute (ILRI), pp. 47.

Vaca Diez, R. (2002). Diagnostico e Impacto Socio-Economico de la Produccion Lechera en Santa Cruz. Su Potencial y Políticas de Apoyo. Memoria Cuarto Simposia de Productividad Lechera 7-8 de Noviembre 2002. UAGRM, Santa Cruz, Bolivia, pp. 133-141.

Index

ND - #0141 - 170726 - C0 - 240/170/18 - PB - 9781789183047 - Matt Lamination